PRAISE FOR THE GRASSFED GOURMET FIRES IT UP!

"All the conscientious carnivore will ever need: a slab of grassfed beef, a blazing hot grill, and this mouthwatering book."

—Nina Planck, *The Farmer's Market Cookbook*, *Real Food: What to Eat and Why*, and *Real Food for Mother and Baby*

"A beautiful harmony from this chef-farmer duet. Heller's philosophical and explanatory grass-farming ruminations bring pastoral delight onto Calvert's grill and into your kitchen. What a fabulous addition to pasture-based culinary expression."

—Joel Salatin, Polyface Farm and author of *The Sheer Ecstasy of Being a Lunatic Farmer* and *Everything I Want to Do Is Illegal: War Stories from the Local Food Front*

"A sense of passion and purpose animate this work with its far-reaching range of information on the grassfed movement and the delights of preparing meat full of flavor and healthful nutrients. These 'grassfed gourmets' take the reader on a journey from the tranquil grassland of America, to no-nonsense, how-to techniques of traditional smoking and grilling; culminating with the best collection of mouth-watering grassfed recipes imaginable."

—John Shields, chef, television host, and author of *Coastal Cooking with John Shields*

"With its detailed description of pasture-raised animals and grass-based farming, *The Grassfed Gourmet Fires It Up!* is not just a grilling cookbook, it's a valuable educational tool that happens to be filled with delicious, mouthwatering recipes and a must read for anyone who fires up the coals."

—Diane Hatz, the Glynwood Institute for Sustainable Food and Farming

"It's not just the marinades, rubs, and brines and sauces—that's all here and more—it's the marriage of great information with some hard-earned philosophy that makes this book as lush as the pastures that nourish the animals it describes. Read and you'll never look at a cow or goat or steak or roast the same way. Enlightening and mouthwatering, even the most hard-core vegetarian might reconsider their ways."

—Michael Ableman, author of *From the Good Earth, On Good Land*, and *Fields of Plenty*

"Today's close working relationship between farmers and chefs not only contributes to everyone's eating pleasure, it provides a wealth of information that we can use in our own farms and kitchens. In this lovely book we learn how to produce and prepare meat from grassfed ruminants. Enjoy!"

—Frederick Kirschenmann, president, Stone Barns Center for Food and Agriculture

"Chefs and consumers are turning their attention to the benefits of livestock raised on grass, and not a moment too soon. Whether one looks at health benefits, environmental advantages, or animal welfare, grassfed is best. Now there's a hunger for information about where to find it and how to cook it. *The Grassfed Gourmet Fires it Up!* is a great addition to the library of anyone interested in joining the grassfed revolution."

—Bill Niman, founder, Niman Ranch and BN Ranch

"*The Grassfed Gourmet Fires It Up!* is a must read for anyone interested in how we produce our food, moving toward a more sustainable agriculture and how somone committed to ecologic agriculture and sustainable food systems can enjoy eating grassfed meats. Rita Calvert and Michael Heller do a great job of highlighting many of these connections and the multiple benefits of 'pasture-based' farming practices and food preparation. These lessons are key to understanding how we can move away from today's unsustainable industrial food animal production system—and toward a more sustainable and resilient ecologic animal agriculture that embraces what Aldo Leopold called the land ethic."

—Robert S. Lawrence, MD, director,
Johns Hopkins Center for a Livable Future

THE ULTIMATE GUIDE TO GRILLING, SMOKING, AND BARBECUING PASTURE-RAISED MEATS FROM AMERICA'S SMALL FARMS

Rita Calvert and Michael Heller

FOREWORD BY **ALIZA GREEN**

EATING FRESH PUBLICATIONS

HAVERFORD, PA

About Eating Fresh Publications

Eating Fresh Publications seeks to connect consumers to local and sustainable agriculture and to promote the taste, health, and community benefits of eating foods that are grown and raised the way nature intended. Through cookbooks, guides, and events, Eating Fresh demonstrates its support for local food systems, sustainable farming, the humane treatment of animals, and the transformation of the way we shop for, cook, and think about food.

410 Lancaster Ave.
Suites 4 and 6
Haverford, PA 19041
609-466-1700
www.eatingfresh.com

Published by Eating Fresh Publications

Cover photography: Rita Calvert and Joanna Tully
Authors photograph: Joanna Tully
Interior photographs: Rita Calvert, Wendy Rickard, and Joanna Tully
Cover design: *the*BookDesigners
Design and layout: Margaret Trejo
Recipe editor: Gerry Gould
Indexer: Lee Lawton

All recipes in this book have been tested for home preparation.

Printed in the United States on recycled paper.
First edition
ISBN: 978-0-9673670-4-0
Library of Congress Control Number: 2011921214

To all the dedicated farmer and grower friends who taught me about the world of sustainable living
—RC

To the fabulous farmers and members of the Maryland Grazers Network; and to my daughter Jessie (who knows why)
—MH

also available from
eating fresh publications

The Grassfed Gourmet Cookbook: Healthy Cooking and Good Living with Pasture-Raised Foods

The Great News About Grass (pamphlet)

Contents

Foreword

by Aliza Green

Growing up in a kosher home, where the choice of what to eat is always mindful and follows the biblical mandate that one must not cause pain or suffering to any living creature, I learned that every animal must be fed and watered, then slaughtered quickly using an ultra-sharp knife. Today, encouraged by what I've learned in *The Grassfed Gourmet Fires It Up!*, I am trying to eat meat from animals that have been fairly treated, serve meat in smaller portions, and minimize waste by saving trimmings to make soup stock. I'm learning to work with grassfed meat, which has much different characteristics than either the tasty-but-tough kosher forequarter cuts that I grew up eating or the richly fatty grain-finished beef that I served as a chef to my customers.

After meeting and speaking to Temple Grandin, world-renowned expert on animal behavior, I learned that, unfortunately, both kosher and conventional packinghouses alike slaughter far too many animals in far too short amounts of time and space. Some countries, including New Zealand, Norway, and Switzerland, ban even kosher slaughtering because the animals are not permitted to be stunned beforehand, which would also apply to the Halal meats eaten by Muslims. Whatever our religious beliefs, we all need to treat animals well during their lifetimes, respect that they have sacrificed their lives for us, and not be greedy in using too much of the world's resources by consuming large quantities of meat.

No matter what the animal, the story is always the same: too many creatures living in one place makes for huge health and food-safety issues. Too many pigs means giant lagoons of waste that can leak into surrounding soil; too many cows and antibiotics must be used to prevent illness, thereby creating other problems; and too many chickens and we get salmonella. A single packinghouse can butcher as many as 32,000 pigs in a single day. Nature is telling us that animals need enough room to move around, that they need to eat a natural diet adapted to their particular bodies, and that care must be taken to keep them calm before slaughter.

While researching my *Field Guide to Meat*, I attended a three-day intensive course at Texas A&M University called "From the Stockyard to the Table." Out of 32 attendees, only one, a woman, owned a grass-based ranch. Most of the others worked for one of the four big packing houses (read slaughterhouses) known as "confined animal feeding operations" where up to 450,000 cattle may be penned in at one time in giant feedlots. Today, those four companies alone account for more than 70 percent of the 35,000,000 cattle harvested annually in the United States. About 98 percent of beef at supermarkets comes from those cattle, which are fattened on grain, especially corn, for three to six months before slaughter.

Much of the work that local butchers used to do (breaking down quarter or half-carcasses into smaller wholesale and retail cuts) is now done at the packinghouse itself. This cuts costs because the least valuable parts—fat, bones, and trim—are trimmed before shipping, but it also creates a further disconnect between the between consumer, animal, and farmer. Along with the growth in sales of grassfed and pastured meats, we are seeing a new cadre of proud artisanal butchers who have the knowledge and skill to make the most of the meat they sell.

Because of increasing demand by consumers for meats produced locally and ethically from places like Michael Heller's Clagett Farm in Maryland, the numbers of smaller-scale meat processors are increasing, especially those servicing grassfed operations. Some of them are even mobile, traveling from farm to farm. One processor, True & Essential Farm-Raised Local Meat in Virginia, brings schoolchildren to visit—the ultimate in transparency. Hardwick Beef processes and distributes organic, dry-aged grassfed beef raised by farmers in Maine, Massachusetts, New York, and Vermont to food co-ops, specialty stores, and local food-centric restaurants like Blue Hill at the Stone Barns Center for Food and Agriculture.

In America, we live in a culture of abundance where more is always better and steaks are so big, they hang off the sides of a plate. Recently, *Philadelphia Magazine* did a cover story on the city's astounding 35 major steakhouses. Not only are the menus at these steakhouses all cut from the same cloth, their proliferation points a glaring spotlight on our need to stop overindulging in steaks where size is the sizzle. Instead, let us seek out the grassfed and pastured meats featured in *The Grassfed Gourmet Fires It Up!,* meats that we can feel good about eating. We can make that meat count by preparing it using the rubs, marinades, and sauces created by Rita Calvert for this book.

Grass-finished beef comes from cattle that forage naturally on grasses and legumes their entire lives. The quality of their forage is so important that their ranchers often refer to themselves as "grass farmers." These cattle live low-stress lives and they are not fed growth hormones, so they are generally healthier, without the use of antibiotics. Their meat is higher in vitamin E, beta-carotene, vitamin C, and the "good fats": omega-3 fatty acids and CLA (conjugated linoleic acid). The cattle also fertilize the land in amounts that the soil can safely absorb. However, the process of raising cattle on pasture is time consuming, making the end product more expensive than conventionally raised meats.

Grassfed meats can have a bold herbal or gamy taste that can be disconcerting if it's not what you're expecting. Those who love it describe grassfed beef as having a vibrant, bold, and complex flavor that is very different than the flavor of corn-fattened beef that Americans have become accustomed to eating. The texture of grassfed beef is different as well—juicier, not as dense, and less fatty. Without the benefit of moisture imparted by the fat contained in grain-finished meats, these meats can turn from rare to well-done in a matter of minutes.

In beef-loving Argentina and Brazil cattle are grassfed and their meat is served in churrascarias, or steakhouses. They typically offer cuts that Americans tend to neglect, such as the coulotte, or top sirloin cap, known as picanha; the bottom round, known as the fraldinha; and even the plate, the cheapest most flavorful cut of all—equivalent to pork belly. Also, cooking larger cuts of meat, which is how it is done in the churrascarias, will always yield more than smaller, individual steaks where shrinkage becomes significant.

Count on serving at most one-quarter pound of meat per person rather than a typical restaurant steak portion of 12 ounces for a strip steak and 8 ounces for a tenderloin fillet. Buy that 1-inch thick strip steak but serve it for two. Invest in a good knife, and thinly slice steaks like tri-tip, shoulder tender, flank, and skirt steak in the kitchen, fanning these chewier but very tasty cuts out on the plate to make portions bigger to the eye and easier to eat.

Whether you buy a portion of a whole grassfed or pastured animal through a buying group or CSA (Community Supported Agriculture) or you buy retail cuts at a specialty market, the challenge is to use as many parts of the animal as possible. Only 10 to 12 percent of the meat from beef cattle represents the high-priced, tender middle cuts from the rib and loin sections, such as rib eye, strip steak, and tenderloin. And, 50 percent of all beef still ends up as ground meat. Therefore, butchers and consumers have a great deal of often tougher but more flavorful cuts to use in ways beyond grinding them up for hamburger, which fetches the lowest price. No worries, though: in *The Grassfed Gourmet Fires It Up!* Rita Calvert shows you just how to master techniques like barbecuing, grilling, and smoking that work best for a large array of grassfed meat cuts.

With more experience, farmers like Michael Heller are helping raise the standard for grassfed beef so that quality, tenderness, and consistency are improving. With the help of knowledgeable chefs like Rita Calvert, cooks are learning how to adjust their recipes by cooking the meat less and moistening them with fats like olive oil and vegetable oil or even wrapping a lean cut in bacon or pancetta. Her lively and imaginative recipes include some of my favorite flavorful cuts of beef: the small but beautiful shoulder tender, the lean but tasty flank steak, the dense and delicious tri-tip and the brisket, the king of slow-cooked Texas-style barbecue.

As we become more knowledgeable about the meat we eat and how it gets to our tables, the grassfed movement will continue to burgeon. With the help of dedicated "grass farmers" like Michael Heller and local food advocates like Rita Calvert, we gain skill in getting the best results from meats and poultry that we can feel good about eating and take pride in how our choices positively impact the world around us.

Acknowledgments

My utmost appreciation goes to my grandparents, Arlo Bell and Horace, who taught me the essence and art of real food and to savor the freshly plucked scents and flavors from our own garden and nearby farmers market.

Many thanks also to my fellow author and mentor, Michael Heller, who educated me not only about the sustainable farming life, but social justice as well. Our friendly visits with his cows have always been a highlight of working with Clagett Farm, both before and during our collaboration on this book.

Heartfelt gratitude to the Chesapeake Bay Foundation (CFB), which, since 1983, has been fighting and educating to save and restore the largest estuary in North America. This is the waterway I live beside, escape to, and feature in all my entertaining.

Many friends at CBF have guided my path, linking the bay's future vitality to our choices today about agriculture, sustainably raised food, and cooking.

This book would not have been possible without my editor, Gerry Gould. Her impeccable eagle eye assures that our recipes can be followed and understood by chef and novice cook alike.

Praise be to Nicholas and Robbin (my sister) Kuchova for their enthusiasm and keen palates and their willingness to test endless meat and poultry recipes, even while weathering the challenging storms of Robbin's successful treatments for cancer.

To Ann Wilder, no longer with us on this plane, yet continually perched on my shoulder as I wrote about the world of herbs and spices for this book. And always warm hugs to friends Greg Strella, head farmer of Great Kids Farm in Catonsville, Maryland; Jeanne Dietz-Band, geneticist, and founder of Many Rocks Goat Farm; Wendy Child; and all of the farmers who made their excellent products and wisdom available for the making of this very special book.

Finally, I would like to express my heartfelt appreciation to the farmers whose generous donations made it possible for us to learn more about how to prepare these wonderful meats and poultry. Special thanks to Burgundy Pasture Beef, Drover Hill Farm, Evermore Farm, Grassland Beef/U.S. Wellness Meats, Gunpowder Bison and Trading, Hardwick Beef, Jamison Lamb, Lindenhof Farm, Many Rocks Farm, and Polyface Inc.

—RC

Many thanks to the farmers and members of the Maryland Grazers Network for sharing their considerable wisdom; and to the Chesapeake Bay Foundation (a wonderful environmental organization) for understanding (and supporting) the idea that sustainable farmers and thoughtful consumers are key to a healthy Chesapeake Bay.

—MH

Introduction

Almost nothing tastes as good as food cooked outdoors over an open fire. It is even better when you are cooking with meat that comes from animals raised the way nature intended. Outdoor cooking, whether it is grilling, barbecuing, or smoking, is both an art and a science and it is easily mastered by following a few basic pointers.

There was a time when grilling was the primary means of food preparation. Today, we have the luxury of revisiting the cooking techniques of our parents, grandparents, and great-grandparents with the benefit of years of refinement and a lot more information. If they only knew what we know now! How *did* they get by without a digital instant-read thermometer? How did they fare without the ease and consistency of a gas grill or even a small charcoal-burning grill that you can roll around the backyard?

One of the great joys of cooking is that it can send us along a path to discovery. With grassfed and pasture-raised foods, the path to discovery is especially gratifying as we marry the ways in which animals used to be raised (before feedlots) with modern-day cooking techniques.

What does it mean for meats, poultry, and dairy products to be grassfed or pasture-raised? While there are ongoing discussions about the precise definition of those terms, quite simply, animals that eat what they find in the wild are referred to as grassfed or pasture-raised. Depending on the animal, this may mean grass, worms, insects, and other nutritionally rich food sources.

A new culinary approach to cooking—and especially to grilling—is necessary with grassfed and pasture-raised meats. The number one rule for cooking pastured meat is to not overcook it. Grassfed beef needs about 30 percent less cooking time than conventionally raised beef, and it works best when cooked medium-rare to medium. Pastured poultry also requires reduced cooking time. We recommend making a good meat thermometer your new best friend.

Let's reinforce that rule: grassfed meats require lower cooking temperatures and shorter cooking times. We believe all forms of cooking should be done mindfully. For instance, when cooked on the grill, some cuts of meat require close observation to achieve a golden caramelized crust on the outside and a moist, juicy interior. Other cuts benefit from longer cooking times at low temperatures with a bit of liquid, which can be accomplished on the grill as long as one understands the technique.

In the bigger picture, *grassfed* is about more than just cooking. It rapidly is becoming a lifestyle that acknowledges the impact of our food choices. Some people come by those choices naturally; others discover the need for change for specific reasons, such as because they are dealing with health issues and reevaluating their diets, because they are concerned about the environment, or because they saw something on television or on the Internet about the horrors of factory farming.

Moving toward a system of agriculture that favors pasture-based farming has multiple advantages. In addition to providing foods that are more nutritious, pasture-based farming benefits the animals, the farm workers, wildlife, and the environment. When cooked correctly, grassfed meat is healthier and more delicious than meats from animals subjected to feedlots and raised on factory farms. Food from cows, goats, lambs, pigs, and poultry that are raised on pastures is entirely different from what we get from animals raised on factory farms or in feedlots.

What makes it different? Animals raised on lush yet carefully managed fields move freely, exercise, interact with their companions, and eat only what nature intended. The foods they eat suit their unique digestive systems. The ability to move freely prevents unhealthy fat from forming as it does in sedentary animals. Eating the right foods yields leaner, healthier animals than do grains, corn, and even leftover bagels and potato chips, which are common practices in some feedlots. And while health experts have long been telling us to avoid or limit our intake of fats, the fats from animals raised right are healthy fats. You don't need to fear them the way you do the fats from present-day, conventionally raised animals.

It may surprise you that cattle raised on pasture are not fed corn and grains. Those types of feed, which became standard in American agriculture after World War II, are toxic to a cow's digestive system; they are used to fatten cows as well as to produce consistent flavors. Industrial meat production may have made meat less expensive and more accessible, but it harms the animals; it is so detrimental, in fact, that animals contained in feedlots must be pumped with antibiotics to keep them from contracting diseases. Not all grain-fed animals are raised in the industrial model; however, unless you know where your meat comes from, you can be almost certain that the animals were raised in large, industrialized, confinement operations.

We believe *clean* best describes grass-based farming: clean and simple. The animals are clean, their flavors are clean, and the environment in which they are raised is clean and self-sustaining. We all have the power to create a better food system and a cleaner environment, and, as farmer and chef we can't think of a better way to do it.

As we describe the benefits of grassfed meats in this book, we want to stay real. Some claims made about the benefits of grassfed meats are difficult to substantiate, so we believe it best to stick with what we know because what we know still makes a compelling and exciting case for choosing grassfed. For many people, the taste alone is reason enough. Farmers raising grassfed animals use many adjectives to describe the distinct flavors of their grassfed meats: sweet, earthy, rich, and nutty, to name a few. We often hear the textures of these meats described as having integrity, as opposed to the bland softness that producers of industrial-raised conventional meats intend. When it comes to health benefits, most grassfed and pasture-raised meats are clearly lower in saturated fats, and yield a healthy balance of omega-6 to omega-3 fatty acids. In short, eating grassfed meat is like eating fish, but with an appeal distinctly its own.

Whether you have already discovered the pleasures of grassfed or you are just beginning to explore pasture-raised foods, you probably have questions. Here is a list of some of the interesting questions that we hear most.

- What does it mean to raise animals solely on pasture, and why are so many farmers choosing that route (see page 2)?
- What are the health benefits of pasture-raised meats (see page 8)?
- Are there differences in the benefits between pasture-raised ruminants (beef, bison, lamb, and goat) and pasture-raised nonruminants (chicken, turkey, pork)? (Yes, there are! See page 4.)
- How does choosing pasture-raised meats benefit our environment (see page 10)?
- Why would a grilling cookbook advocate eating less meat (see page 11)?

The answers to these questions and so many more are available in these pages. For home cooks, we explore how to prepare various cuts of grassfed meats for maximum flavor, but we also take this opportunity to step back and examine how our actions connect us to the environment and to each other.

So, welcome to our world. We encourage you to get out to a nearby farm or a local farm market, or spend an hour or two in the country, buy what looks good, and dust off that grill. Let's start cooking.

Rita Calvert and Michael Heller

About the Authors: The Chef and the Farmer

Throughout this book, you will hear from Rita, a chef, food educator, and passionate supporter of the local-foods movement, and Michael, a farmer who has been raising beef on pasture for more than 25 years. Our goal is to weave together the dual strands of farm and food, from the fields to the table. We believe that even with the finest—and healthiest—ingredients, it is impossible to fully appreciate the food we eat without knowing about the farms on which it is raised.

RITA'S STORY

The smoky grilled beef and caramelized turkey aromas permeate the air as a breeze rustles the grasses along the water's edge. The ospreys calling from the docks are disrupted by the kayakers who are joining us for the grillfest. The out-of-towners arriving by car remark on the luxury of such a peaceful haven amid the bustling bayside town. I announce the meal is ready and we gather 'round the picnic table for a sampling of the finest 100 percent grassfed beef sliders this side of the Mississippi. The *River Dinner Meteorite Turkey* (page 128), hot off the fire, raises excitement as much for the story behind its name as for the taste of its juicy meat.

A longtime passionate, professional cook, and avid kitchen gardener, I love to entertain. When I do, I find that just a passing reference to "buying local" is enough to get my guests thinking and talking about the sourcing of the food they eat and serve; in fact, they often have shopping adventures of their own to share. At my River Dinners (or breakfast, or lunch, or hors d'oeuvres, or other treats), the conversation invariably drifts off into the unique characteristics of outdoor cooking, the diverse flavors of seasonal and locally sourced ingredients, and the global cultures that influence our cooking.

In the 1980s, I moved from the Mid-Atlantic to Northern California to explore a lifestyle I imagined would suit me. I worked my way up from the bottom of the food-world chain, became a professional chef, and then joined the New American Food revolution. My culinary business partner was an assistant to the legendary James Beard, whose outdoor grilling, especially with meat, became a crown jewel in his vast repertoire. I was fortunate to have the bounty of the West Coast at my doorstep. Eventually, I added my signature to a restaurant, Cedar Street Café, in Santa Cruz. The café customers became my guinea pigs, eager to test my mesquite-grilled menu from A to Z. The world of live-fire cooking with nutritious, locally grown and raised ingredients became my playground. I even had a kitchen garden in the backyard of the cafe—complete with sorrel and borage, two of my favorite herbs. I launched an award-winning line of mustards as a marketing tool for the café and created a motto: "Bumpy Beer Mustard with a keg of beer in every batch." Those condiments familiarized me with food manufacturing and what happens when food is produced in large quantities. I was a small-volume manufacturer and, eventually, I came to understand the difficulties manufacturers face when trying to make mustard packaged in a glass jar affordable.

Several years later, the call of family brought me back to the Mid-Atlantic. I settled in Annapolis, Maryland, and saw with fresh eyes the ripeness and beauty of the region—so close to the nation's capital—and how easily it is every bit supportive of fresh, seasonal food from small, high-quality farms, as California. Farmers like Michael Heller are at the forefront of the local foods movement and I am honored to collaborate with him.

Attitudes toward food and farming may be evolving here in the East, but nothing has changed so dramatically in the past three decades as the food world in general. More than ever, culinary professionals are playing active roles in highlighting the superior taste and exceptional quality of locally grown and raised foods. Today I can see how my culinary adventures have come together: I delight in helping build bridges from small artisanal farms to tables through cooking classes, food writing, and, more recently, photography.

The Chesapeake Bay Foundation (CFB) is the great unifying force between Michael and me. It is where our lives in agriculture and the culinary arts converge. Out of that connection, I became a volunteer with CBF so I could help more people make the connection between food, farming, and the health of the Chesapeake Bay. The joy of doing so, as well as the good fortune of working with Michael and Clagett Farm, have bolstered my commitment to spreading the good-food message.

I am forever fascinated by the beauty and simplicity of sustainably produced food, including the animals that live healthy and contented lives on pastures. I suspect my interest is in my genes because my mentor—my grandfather—grew up on a farm. I feel his farm upbringing in my roots when I visit friendly teaching farms such as Clagett that have become my foundation and my inspiration.

MICHAEL'S STORY

Rita and I met several years ago, here at Clagett Farm in Upper Marlboro, Maryland. Rita had called me, brimming with excitement about local foods, wanting to discuss the possibilities. When she arrived, I happened to be mending a pasture waterline that I had accidentally clipped with the hay mower. Sitting in the persimmon tree pasture while tightening hose clamps, we talked of ways to connect consumers and local farmers. Rita has the energy and enthusiasm of at least two people, and she has created a vortex of activity around local foods here in the Mid-Atlantic. Over the years, we have supported each other's many local-food-and-farm stewardship efforts. Working together on this book has been delightful. We each offered our very different experiences and perspectives. And, we both have thoroughly enjoyed sharing our passion for grassfed grilling and the challenge of strengthening the local food systems that link farmers and eaters.

It would be great if you could join me for an early morning ramble on the farm. After a steaming cup of coffee, we would walk out across the fields to move the cows to fresh pasture—my daily start to farm chores. If it happened to be a misty morning, and I was feeling in the mood, I would bring my bagpipes. The cows are my greatest bagpipe fans (perhaps my only fans) and they gather around me, transfixed by the haunting sound of the pipes on a foggy morn. There is no question that bagpipes have inspired many a joke, and my friends have mined a rich vein of them from the cows' fascination with my pipe playing: "They're just checking to see if it's an injured animal" or "They're really protecting their young." However, I can say with some confidence that the cows and their calves find the sound of the pipes quite mesmerizing, perhaps carrying them back to their ancestral roots. They are, after, all British breeds (Red Angus, Red Devon, Shorthorn) with strong ties to the British Isles.

My story revolves around the things I love most: my family, farming, and the environment and nature. For me, these three loves are as hitched together and integral to each other as a team of good draft horses. Raising three kids on the farm with my wife Helen, a schoolteacher, has been rewarding. An old farmer once said, "The farm is the portrait of the farmer." In the same way, I believe that farm kids create a portrait of the farm. When I look at my kids—their values, interests, and sense of fun—I see our farmland and animals reflected in them. Growing up, the kids would all help out—baling hay with me, trimming umbilical cords of the newborn calves, and even taking over farm chores when I needed to leave for a few days.

I remember one late winter when I had to leave the farm for a week. Our oldest child, Jessie, who was nine at the time, took over feeding the cows and keeping an eye on the cows that were calving. Jessie had to walk the half-mile to the barn twice a day through snow a foot deep. Three cows calved while I was away; her face glowed with pride when she showed me the new calves upon my return.

For 27 years, I've worked the hilly, 285 acres of Clagett Farm, which is owned by the Chesapeake Bay Foundation (CBF). In helping with agricultural and environmental policy issues for CBF, I have been able to connect my love of farming with the broader goal of caring for the environment. It has been a tremendous opportunity to interact with diverse, interesting individuals, discussing the impacts and benefits of different farming systems.

I have many vivid memories of spending quality time with delightful and interesting farmers and others at farm meetings. At one meeting in Arkansas one of the "extracurricular" activities that I instigated was to assess the water quality of the White River firsthand. José, Billy, Ricardo, Freddy, and I went to the local hardware store for a strong rope, and then selected a likely looking dirt back road down to the river. We made an amazing rope swing and used the levee as our platform for takeoff into the river—getting a wonderful sampling of the White River.

At another meeting I learned a valuable food and farming tidbit about "sheep dip" while lunching with some local farmers at a pub in Kent in southern England. As we were eating, Tom Bryson, an old sheep farmer in the group, said, "We need to introduce Michael to sheep dip." They all chuckled as they saw my eyebrows crinkle in bemusement. They knew I knew that sheep dip is a toxic drench used to soak sheep to kill external parasites. They also knew I *didn't* know that sheep dip had taken on another meaning by farm hands and feed stores. Often a farm hand, when picking up farm supplies, would ask the store clerk to toss in a bottle of whisky and put it down on the landlord's bill as sheep dip. Sheep Dip is now a brand name for a rather distinctive whisky.

My role with CBF of connecting the farm and environmental communities enabled me to be a founding member of Future Harvest, which has been Maryland's sustainable farming organization for the past 12 years. More recently, working with CBF and other organizations, I helped start a new mentoring group, the Maryland Grazers Network. The Grazers Network is a tremendous collaboration of farmers (beef, dairy, goats, and sheep) working to help new and younger farmers who are interested in becoming really good *grazers* (grass farmers). I've probably gained more practical knowledge from these farmers than from any other group. (For instance, from one of our delightful Mennonite dairy farmers we all learned that you should "cut thistles after the strawberry moon" to eliminate them from your pastures long term.) Amongst the flotsam and jetsam of ideas in this book are many treasures washed ashore from their collective wisdom. Any earnings from publishing *The Grassfed Gourmet Fires It Up!* that I receive go straight to CBF and the Maryland Grazers Network because I know they are doing truly important work on the ground every day.

If you can imagine accompanying me through the pastures some morning, our boots wet with fresh dew, you will see cows peacefully grazing rolling green pastures. The calm and simplicity of the scene belies the wonderful ecological complexity of the fields and the interaction of cows and pastures. Understanding this complexity and the benefits of farmers working with the natural ecology are what I hope I have brought to this cookbook.

Because of my interests and loves, working on this book has been more than just a creative and professional venture; it has been a way of sharing the things I find most fascinating and important. You might say that this project got my attention much the way the echo of a farmer's bagpipes up on a hillside catches the notice of a herd of cows on a misty morning.

1

A Farmer's Take on Grassfed

by Michael Heller

The term *terroir* is of French heritage. It stems from the phrase *gout de terroir* or, literally, *taste of earth*. It implies that when produced well, a food or beverage possesses unique attributes that represent a place. In the United States, *terroir* appeared in our lexicon by way of the highbrow wine trade. These days, it describes coffee, cacao beans, raw milk, artisan cheeses, cigars, and now grassfed meats. Even if you regard the term as lofty, the idea of terroir does make sense. You can ask farmers about it. They probably will not use French to describe the influence of an animal's environment on the taste of its meat, but that's really only semantics.

It should come as no surprise that the physical and environmental characteristics in and around a particular crop or animal, such as climate, soil composition, and geographical location (including slope), are imparted to the final product. We just don't often take the time to think about it. Rita recently visited an alpaca breeding and fleece farm. The owner, who is quite the educator, described the influence that various pasture grasses have on the quality of the wool. He constantly monitors the addition of alfalfa and other pasture grasses because a precise balance protects the correlation between the grass (nutrition) and the condition of the coat, which can change quickly.

To farmers, terroir is really just the down-home flavor of our meats. It is the unique combination of growing conditions we experience on the farm. A rib eye steak tastes different depending on whether it came from Billy's farm or Ryan's farm. It all starts with the soil, which reflects the farmer's land and the weather. For instance, Billy's farm tends to get more rain than most as more storms seem to track down along Maryland's Patuxent River. This creates a *microclimate* that shapes the soil on his farm, which, in turn, determines the variety and quality of plants in his pastures. The subtle differences between farms result from the smorgasbord of assorted plants we provide for our animals, the breeds of animals we prefer to work with, and even our different temperaments as farmers.

In stark contrast to pasture-raised meats, which celebrate terroir and have wonderful, individualized traits, conventionally raised meats are, by design, the opposite of down-home. They are intentionally anti-terroir. The goal of the industrialized meat-production system is uniformity and consistency. Imagine yourself eating in fast-food restaurants in which each item on the menu tastes the same no matter where you are in the world. Industrialized meats taste the same whether the animals were raised in California or Nebraska or North Carolina. The only way to achieve uniformity is to raise animals in confinement where the feed and the growing environment are totally controlled. The flavor of grassfed meats is more complex, influenced by local soils and by the seasonal differences of the plants, and whether the farmers' animals were harvested in the spring, fall, summer, or winter.

This all speaks to the important and delightful benefits of getting to know farmers and their farms firsthand when you can. Grass farmers thoroughly enjoy showing city folks around!

THE WHAT, HOW, AND WHY OF PASTURE-RAISED MEATS

Grassfed,* *grass-finished*, and *pasture-raised* are all terms used to describe meats produced from animals raised as nature intended—in the open, on pastures, and without corn and soybeans to fatten them. In November 2007, the U.S. Department of Agriculture (USDA) published a voluntary standard for grassfed that specifies, "grass and/or forage shall be the feed source consumed for the lifetime of the ruminant animal, with the exception of milk consumed prior to weaning. The diet shall be derived solely from forage and animals cannot be fed grain or grain by-products and must have continuous access to pasture during the growing season."

Although the government has taken an interest in standardizing grassfed for marketing purposes (and to help consumers understand what the USDA meant by *grassfed*), the terms used to describe pasture-based farming can have different meanings for different farmers. Therefore, it is important to know the farmers you deal with and to

---

* Editor's note: After several of our manuscript reviewers asked why the term *grassfed* is not being hyphenated, we thought it was worthy of an explanation. While we generally wait until compound words are recognized as single words by the dictionary before doing the same, we decided that the term grassfed has become common enough, particularly in the sustainable agriculture and culinary worlds, to warrant treatment as a single word.

understand exactly how they raise their animals. That's the only way you can be certain of the nature and quality of the meats you purchase. (In Getting to Know Your Farmer, on page 35, we spell out some of the points we discuss when we purchase meats from another farmer.)

Saying that pasture-raised animals are raised "as nature intended" requires an explanation. We should make the distinction between ruminants and nonruminants to highlight the important differences between meats from animals raised on pasture. For starters, ruminants include cows, sheep, bison, and goats (there are others such as deer, alpacas, and elk). The ruminants can get 100 percent of their diet from pastures alone. The nonruminants we include in this book are chicken, turkey, and pork. These animals cannot survive on grass alone; their diets need to be supplemented with grain or fruits and vegetables. The dietary variations between ruminants and nonruminants lead to important differences, particularly regarding issues like global warming and nutrition. Thus, for our purposes, we use the term *grassfed* to refer to ruminant animals, and *pasture raised* in a general sense to include ruminants and nonruminants (pastured poultry and pastured pork). Some farmers who work with a multitude of species may graze beef cattle, goats, and chickens, or some other combination of ruminants, nonruminants, or both. The numerous benefits of raising animals on pasture are especially significant regarding ruminants, but most benefits apply to all the animals included in this book.

What is grassfed?

Whether it is cattle or chickens, pigs or bison, animals are meant to eat what they find in the wild. When animals are able to eat what they find on lush, green grass, the foods they produce are referred to as "grassfed" or "pasture raised." In 2007, the U.S. Department of Agriculture issued a voluntary standard for grassfed that recommends that grass and/or forage be the sole feed source consumed for the lifetime of the ruminant, with the exception of milk consumed prior to weaning. The American Grassfed Association (AGA) defines grassfed as food from ruminants (including cattle, bison, goats, and sheep) that have been fed nothing but their mother's milk and fresh grass or grass-type hay from birth to harvest and food from nonruminants (including pigs and poultry) that are fed grass as a significant, but not exclusive, part of their diets, since these animals need to consume some grains. Farms and ranches that meet the AGA's standards may qualify for certification. The AGA's standards are verified by auditors from Animal Welfare Approved (www.animalwelfareapproved.org) (see page 5).

Ruminants: Walking Breweries

Cows and other ruminants (sheep, goats, and bison) grow magnificently on grass because they are walking mini breweries. They have evolved large fermentation vats in their stomachs so that they can digest the energy contained in the leaves and stems of grasses and shrubs. For other large animals and for humans, almost half the stems and leaves of plants are totally indigestible because about 50 percent of stems and leaves comprise cellulose; our digestive systems have never developed the necessary enzymes to break down cellulose into usable energy. Well, neither have ruminants, but they have come up with an ingenious solution to digesting cellulose: they enlist billions of tiny helpers.

I remember as kids we'd say that cows have four stomachs. Later, of course, I learned this really isn't true, although it's not a bad way to describe the cow's digestive system. A cow's stomach has four compartments: the first three chambers serve as fermentation vats that break down the cellulose for the fourth chamber, which is a lot like the human stomach. We sometimes picture cows lazily chewing their cuds, and it certainly is peaceful to hang out in a pasture with cud-chewing cows while ruminating on our own philosophies. While you daydream, the cows, through chewing their cud, are working: they are mashing the stems and leaves that are then readily fermented in their three forestomachs. The rumen, by far the largest of these three fermentation chambers (hence the name, *ruminants*) contains hundreds of billions of bacteria, protozoa, and yeasts. These bacteria and yeasts have the enzymes that break down the cellulose and release the energy cows need.

What's my point? Ruminants are born to do the tough stuff. Their bodies are distinctly adapted so that they can digest materials other animals—including people—can't. Raising ruminants naturally, heeding their evolution, leads to myriad benefits for the animals, the farmers, the environment, and nutritionally for us.

Raising Grassfed Meats

I've been raising these walking mini breweries (in my case, beef cattle) for more than 25 years, and I am part of a group of farmers we call *grass farmers*. Although grass farmer may not be the most rugged-sounding name, every one of the grass farmers I know takes great pride in that name because of what it embodies. As I mentioned earlier, here in Maryland, we formed a Grazers Network that includes beef farmers, sheep farmers, dairy farmers, and goat farmers . . . every one a grass farmer.

Typically, farmers producing such a range of different animals would not have a whole lot in common, but our group focuses on growing really great pastures (grass), and that's what binds us together. We can share and learn about better and different ways to manage and improve our pastures and raise our animals. When we talk about grass, we are referring to all the edible plants in a pasture (including grasses such as timothy, orchardgrass, bluegrass, and crabgrass as well as broadleaf plants like clovers, dandelions, and lambsquarters) that are nutrition for our animals.

If we let this mixture of plants grow to between 18 and 24 inches, and then cut and dry it in the sun before baling, it becomes the hay that we feed our animals in the winter or during droughts when pasture is scarce. Our goal is to have a healthy diversity of plants (particularly clovers and the true grasses), and to keep these plants in their young and tender vegetative growth stage when they are most palatable and most nutritious for animals. Once plants get tall, brown, and stemmy, they are useless for growing good pasture-raised meats. (Remember, when you visit a farm, to ask yourself: Are the pastures lush and green, or tall and stemmy?) However, there is a new school of grazing that turns some of these ideas on their heads. Called high-density grazing (or mob grazing), it is a strategy whereby large numbers of animals are put on a small area to graze for just a short time—from as little as two hours to up to 24 hours. In this type of grazing some of the fields may have stemmy vegetation with tender vegetation growing below. There is some thinking (and strong anecdotal evidence) that this system may be a step forward in managing pastures. A few farmers in the Maryland Grazers Network are testing out this high-density grazing system and we are carefully tracking results. Stay tuned!

As grass farmers, we think of our cows as primary tools for managing the grasses. Ideally, we let our animals graze in one pasture for only one to two consecutive days. We size the pastures so that the cows will eat everything within that time. We then move the cows to newer pasture and let them graze there until the grasses and clovers in the first pasture have had time to grow back. This system of moving cows from pasture to pasture every few days is often referred to as *rotational grazing*. If cows are not grazed

Animal Welfare Approved

The Animal Welfare Approved third-party certification program and label is regarded by World Society for the Protection of Animals as offering the most rigorous and progressive animal-care requirements in the nation. The pasture- and range-based farms in the program are annually audited by experts in the field, and farms in the program are putting each individual animal's comfort and well-being first. You can find the Animal Welfare Approved label at grocery stores, restaurants, and farmers markets. For more information, see www.animalwelfareapproved.org.

rotationally, they can ruin a pasture. Cows are a lot like children: when left to their own devices, they will eat only certain, favorite plants—the mashed potatoes and gravy of the field—and leave behind less desirable plants—the broccoli and spinach of the field. The plants the cows don't like thrive; they flower, seed, and multiply. Over time, this type of grazing is detrimental to pastures because desirable plants never get a chance to regrow properly, and gradually they die out. We grass farmers don't let our cows become too choosy about what they eat. Instead, we prefer that they eat everything in the pasture before moving on to newer pasture. This makes for healthy and productive pastures that, in turn, grow some great cows. The finest and most tender meats are produced when cows are moved to new pastures daily and allowed to eat the green and tender grasses that provide the highest energy and nutrition.

Conventionally Raised Meats

So, if raising animals on pasture is the preferred option for all types of meats, what is the conventional system that is responsible for most of the meats sold in this country? Well, just about all of the beef cattle, chicken, pigs, and most of the lambs sold in grocery stores and butcher shops (and that are served in restaurants and fast-food chains) are fattened on grains, such as corn and soybeans. Not only are these animals fed lots of grains, but they will also do more traveling than the average American. For example, many beef cows are raised on the hillsides of the lovely Shenandoah Valley of western Virginia. Although many of those cows are born in Virginia, many other young cows have been shipped there from the high plains of Kansas and eastern Montana.

These young cows join their Virginia cousins to be raised on Virginia pastures until they grow to 1,000 pounds or larger; at that point, most will be shipped to places like Nebraska where they are confined in feedlots with thousands of other beef cattle and fattened on grains to roughly 1,400 pounds. They then are sent to Chicago to be harvested, processed, and shipped—as meat—across the country. This elaborate system for raising beef is the typical product of the bizarre economics of fattening cows on grain; that is, it is cheaper to ship the cows than it is to ship the grain. At first this seems counterintuitive, but because it takes approximately five pounds of grain to produce one pound of beef, the economic driver here becomes clear (1).

It is easy to see why it makes sense to work *with* nature to produce beef, and to recognize that natural evolution is responsible for allowing these animals to thrive on a diet of forages, such as grasses and clovers. It is easy to understand the greater sustainability of local pasture-raised meats when compared to the expenses associated with energy costs, the consequences of inhumane treatment of animals, and the environmental problems caused by cramming a thousand or so heads of cattle on a few acres of land.

The Benefits of Pasture-raised Meats

While I consider myself a pragmatist and an ardent environmentalist, as a farmer I have to say that the single most important reason I raise cows on pasture is because it is more humane. Few farming practices pain me more than confining animals on concrete in barns or in *loafing* yards. The number of animals lost to illness and injury combined with the regular use of antibiotics make the entire industrial meat-production system appalling. Even if there were no other benefits to pasture-based farming, just seeing healthy animals out grazing on open fields of well-managed pasture would be reason enough to be a grass farmer.

However, the humane treatment of farm animals is just one of a number of important reasons why we need to promote pasture-based farming practices. All of the benefits are key to understanding the tremendous value of shifting to a diet of pasture-raised meats, and I want to highlight each of them.

Flavorful and Stress-free

Pasture-raised meats are ideal for people who are interested in flavor. Earlier, we discussed the role *terroir* plays in the unique flavors of pasture-raised meats. Unfortunately, most people born in the United States in the last half of the twentieth century have not had the opportunity to experience those flavors. As a result, most people have become accustomed to the marbled meats that are high in the wrong kinds of fats, which develop in grain-fed animals. (A few small farms choose to finish their cattle with grain while managing to raise their animals humanely. They find that some customers remain attached to the traditional marbling in their meat.)

As confinement became more commonplace, the animals were becoming more stressed. Interestingly, stress in an animal's life—throughout its life from birth to harvesting—is one of the key factors in determining the toughness or tenderness of its meat. Animals that lead relatively stress-free lives tend to produce meat that is more tender. The heavy marbling and fat content of grain-fed beef disguises the suffering most of these animals have endured, including long and arduous journeys, confinement, hormone implants, and regular treatments of antibiotics. Once we understand the forces at work, it doesn't take much imagination to figure out that pasture-raised animals, especially those that spend their entire lives on only one farm protected from such extreme stresses, will yield more tender and flavorful meat.

Of course, raising animals on pasture can lead to considerable variations in the taste and tenderness of grassfed meats. That's why flavorful and tender grassfed meats require skilled and experienced farmers. Processing also contributes to the quality of meat. Ideally, meat from grassfed beef, bison, and sheep should be hung to *dry age*, which makes it even more tender. In addition, the shrinkage that occurs during this process helps concentrate and intensify the flavor of the meat. Dry aging is another reason to choose a local farmer with some care.

Nutritional and Health Benefits of Grassfed Meats

We are limiting our discussion of the nutritional benefits of grassfed meats to ruminant animals, although we include nonruminants (pastured poultry and pork) in our recipes. Pasture-raised nonruminants do provide most of the benefits mentioned in this chapter, but they do not provide comparable nutritional benefits to those of ruminants. This is because nonruminants all require some grain or fruits and vegetables in their diets. They cannot get all their nutrition from grasses alone so the term grassfed does not quite describe them. This means that chicken or pork raised in pastures will not necessarily have higher levels of omega-3s or CLAs than the chicken or pork raised in confinement. For chicken or pork to be higher in omega-3s the animals must be fed feeds high in omega-3s, such as flaxseed. Thus, this discussion of nutrition and health benefits of grassfed meats is limited to the ruminants—beef, bison, goat, and lamb.

Humans evolved as hunter-gatherers who survived on a variety of plants and pure, grassfed meats. About 10,000 years ago, with the advent of farm-based agriculture, the human diet underwent significant changes. Nevertheless, the meat portion of our diets remained essentially unchanged until roughly 150 years ago, when it underwent yet another dramatic change (2), one that gravely affected the quality of meats that most North Americans consume today. Much of the meat now consumed in the United States is believed to increase the risks of cardiovascular disease, high blood pressure, and other so-called diseases of civilization (3).

So what happened roughly 150 years ago? In order to shorten production time and increase profits, farmers began feeding their herds substantial amounts of grain. The grassfed meats that our ancestors ate were considerably different nutritionally from 99 percent of the meats now eaten in the United States. Today, beef cattle in feedlots are fattened on grains to the point of obesity (up to 30 percent of their body fat) (4). In less than two centuries, the meat portion of our diet has changed in ways that human evolution could not possibly keep pace with. Therefore, returning to a diet of pasture-raised meats would provide numerous valuable nutritional benefits.

Dietary saturated fat, also known as saturated fat

Grassfed meats contain markedly less saturated fat compared to grain-fed meats. Dr. Susan Duckett of Clemson University reports that grassfed beef has a 53 percent reduction in the bad saturated fats compared with grain-fed beef (5).

Health implications: Nutritionists strongly recommend that people reduce saturated fats in their diets based on the decades of research that have shown high amounts of saturated fat in the diet increases the risk of coronary heart disease (6).

Ratio of omega-6 and omega-3 fatty acids: Essential nutrients

Omega-6 and omega-3 fatty acids are essential nutrients for healthy diets, but it is the balance of those fatty acids in diets that is important. Omega-6s compete with omega-3s, which often impedes the development of omega-3s in the human body, resulting in an imbalance that can lead to health problems.

The ideal dietary balance of omega-6 to omega-3 fatty acids is 1:1 or 2:1. Today, the average North American diet has a ratio of omega-6 to omega-3 between 11:1 and 30:1. With a ratio of 2:1, grassfed meats can help balance diets toward the levels between 1:1 and 4:1 that most nutritionists consider desirable (5). It should be noted that while the balance of omega-6 to omega-3 fatty acids is very positive in grassfed meats, these meats do not have the high levels of omega-3s found in fish.

According to Dr. Loren Cordain of Colorado State University, "The case for increasing omega-3 fatty acids in the U.S. diet has broad and wide sweeping potential to improve human health. Specifically, omega-3 fatty acids and their balance with omega-6 fatty acids play an important role in the prevention and treatment of coronary heart disease, hypertension, type 2 diabetes, arthritis and other inflammatory diseases, autoimmune diseases, and cancer" (4).

Ratio of Omega-6 Fatty Acids to Omega-3 Fatty Acids				
Human evolutionary diet	Healthy human diet	Typical American diet	Pasture-raised beef	Grain-fattened beef
1:1 to 2:1 (7)	1:1 to 4:1 (8)	11:1 to 30:1 (8)	2:1 (4)	10:1 (4)

Health implications: Eating grassfed meats can improve the important balance in the ratio of omega-6 to omega-3 fatty acids in a person's diet. Although chickens do not benefit from grass in terms of significantly increased levels of omega-3 fatty acids, it is worth noting that they can have higher levels of omega-3s if they are fed substantial amounts of flaxseed, which has a 1:3 omega-6 to omega-3 balance.

Conjugated linoleic acids: An anticarcinogenic

The average person may not be familiar with conjugated linoleic acids (CLAs), but health experts say that CLAs can play an important role in strengthening the immune system. On average, on a per-fat weight basis, the concentration of CLA is between two to three times higher in grassfed beef as compared to grain-fed beef (4). In addition, the amount of vaccenic acid is four times higher in grassfed beef. This acid can be desaturated to CLA in the human body (9).

Health implications: According to Susan Duckett and Enrique Pavan of Clemson University, "Conjugated linoleic acids, specifically the cis-9 trans-11 isomer, has been shown to possess anticarcinogenic effects (9). In addition to potentially neutralizing cancer-causing substances, recent studies indicate positive effects of CLA in reducing inflammation, strengthening the human immune system, and lowering risks of heart disease" (6).

Protein and other valuable nutrients

Because grassfed meat is so much leaner than grain-fed meat, a similarly sized portion will provide substantially more protein. It turns out that this lean meat also has higher levels of valuable nutrients. Research accumulated at Colorado State University shows that grassfed meats have a much higher ratio of protein to fat as compared to grain-fed beef: 76.5 percent of its total energy is protein whereas in grain-fed beef only 48.9 percent of its total energy is protein (4).

Health implications: Important trace nutrients—Fe (iron), Zn (zinc), vitamins B12, B6, and niacin—are concentrated in the *lean* muscle tissue of beef, which means that replacing a diet of grain-fed beef with grassfed beef will result in consuming higher levels of these trace nutrients. Also, studies show that grassfed meats have higher levels of cancer-fighting antioxidants such as glutathione and superoxide dismutase, as well as increased levels of precursors for vitamins A and E as compared to grain-fed meats (10). Still, it is important to recognize the benefits of reducing the portion size of meats at your meals. The beauty of eating grassfed meats is that it has been shown that smaller portions are found to be as satisfying as larger portions of grain-fed meats.

Prevention of E. coli contamination

The digestive tract of a grain-fed cow becomes an extremely acid environment, which favors the deadly strains of *E. coli*. Research has shown that even a brief period of switching cows to a grass-based diet alters the environment of the digestive tract so that harmful *E. coli* are less likely to survive and contaminate meats during processing (11). Grassfed meats have a very low risk of contamination with the often deadly *E coli.*

Environmental Benefits of Pasture-raised Meats

Water quality: Clear enough to drink

Common sense (which is not so common these days) will tell you that grassfed animals do most of the feed harvesting (grazing) and the recycling and spreading of many nutrients back onto pastures as manure, which is a good thing. Farmers don't need to plant, spray, fertilize, and harvest corn and other grains and then haul the loads to a feedlot to fatten animals. Corn grown for cattle feed accounts for more than 40 percent of all the commercial fertilizer and herbicides applied to crops in the United States (12). Up to half of that fertilizer will end up polluting surface and groundwater. Also, the nitrogen and phosphorus in the manure concentrated at these feedlots are a critical source of nutrient pollution causing considerable water-quality issues. On the other hand, pastures and hay land can be grown with minimal or no fertilizing inputs; furthermore, because pasture plants tie down the soil, there is no runoff to clog and pollute streams and areas, such as my nearby Chesapeake Bay.

Global warming: Grassfed meat is best, but less meat is better

Regarding global warming, cows and other ruminants present a mixed bag. On the plus side, the plants in pastures and on hay land are great for sequestering carbon by removing carbon dioxide from the atmosphere and transferring it to the soil. In addition, the production of grassfed meats uses considerably less fossil fuel than is required for conventionally raised meats. Because the animals are grazing healthy pastures (harvesting their own feed), farmers don't need to use tractors and combines (which burn diesel fuel) to plant and harvest corn and soybeans, crops that require fertilizers and pesticides made from fossil fuels. The farmers also don't have to take time and use tractors (diesel fuel) to clean manure from the barns and spread it on fields (the cows do this really well!).

However, ruminant animals do produce a large amount of methane gas, a major contributor to the global warming problem. Carbon dioxide gets all the global warming attention, but methane traps more than twenty times as much heat per unit, which makes it the number two gas culprit in global warming. So cows and other ruminants matter. I earlier referred to ruminants as great "walking fermentation vats," and described how this fermentation allows them to digest grass efficiently. Unfortunately, fermentation generates a lot of methane gas and this is a significant environmental problem. But, even in this case, it appears that grassfed animals may provide benefits over conventionally raised farm animals. Some emerging research suggests that grazing and other ways to increase omega-3s in cows' diets (such as including flaxseed) can reduce the methane produced by cows by as much as 20 to 30 percent (13). Another new interesting discovery is the recent finding that oregano may be one of the better ways to reduce global warming! Alexander Hristov, a Pennsylvania State University dairy scientist, has found that when he fed dairy cows an oregano-based feed, it reduced their methane emissions by 40 percent. In a recent conversation with Alex, he said he needs to follow up this one study, but that the best part is that reducing the methane emitted by the cows translated into an increase in milk production of 3 pounds per cow. So it may be that with beef, bison, sheep, and goats a little oregano in their diets will reduce methane emissions and increase the amount of meat (preseasoned of course) produced. Stay tuned.

Wildlife: Grazing provides productive conservation

It always amazes me when I am out baling hay and the buzz of the little grasshopper sparrow reaches me over the throaty roar of the tractor's 4-cylinder diesel Perkins engine. These little sparrows, subtle studies in shades of brown, are one of the many great pleasures I experience on the farm. Unfortunately, they are becoming less and less common as they lose their grassland habitats. *Productive conservation* describes an economically productive farming system that respects and enhances wildlife habitat (grazing, for example). Cornell researchers have found that many bird species, such as the eastern meadowlark and upland sandpipers, have declined by 90 percent in the past 30 years because of the loss of grassland habitats (14). To reverse this trend we need more farmers like Ron Holter, an organic dairy and beef farmer and member of the Maryland Grazers Network, who is interested in the ecology and wildlife of his pastures. Pennsylvania State University researchers found that one of Ron's pastures has 76 different plant species growing. Compare this to the bird and wildlife habitat (or lack of it) you will find in a monoculture, such as a cornfield.

Community and local economics: Make your dollars count

Many people find it difficult to find locally raised grassfed meats, or they prefer the flavor of the meat from a particular farm that is outside of their region. Even though we believe that supporting grass-based farming has important benefits regardless of where you find it, purchasing meats from your local farmer or farmers market has a powerful impact on your local economy. Every dollar you spend with a local farmer will cycle another $2 to $3 worth within the local economy that not only helps save farmers and farmland, but supports the local meat-processing facilities needed to make a local food system successful (15). However, every dollar you spend at a nonlocally owned grocery store or chain supermarket brings only about 20 cents into the local economy! The economist Michael Shuman refers to this as the *multiplier effect*. For a good discussion on the impact of supporting local economies, see Michael's wonderful books, *Going Local* (Routledge, 2000) and *The Small-Mart Revolution* (Berrett-Koehler, 2007).

Humane treatment of animals: If cows could talk . . .

There is no question that if farm animals could talk, they would be the first to tell you they prefer to be out grazing in pastures rather than confined inside a building or feedlot. How do I know? Data show that veterinary and medicine costs are consistently lower for animals raised on pasture (16). The confinement and high-grain diet common in

Costs of pasture-raised meats: paying for quality, not quantity

Sure, there are many aspects of pasture-based farming that are less costly to the farmer (and better for the environment and for the animals). So why, then, aren't those lower costs reflected at the register? There are several reasons. To begin, conventionally raised meats rely on large volume and cheap labor to produce a less-expensive product. Pasture-based meats, on the other hand, come from the hard work of farm families, who raise their animals humanely and who care more about flavor and quality (definitely not quantity) than about volume. Because pasture-based farms do not focus on quantity, they need to charge more per pound in order to make a living. This is another reason to consider buying directly from a local farmer—to reduce your cost and make it possible for the farmers to get full value for their hard work.

---

conventionally raised beef causes dust-related respiratory conditions, metabolic diseases, liver abscesses, and many other health issues (6).

Farmer benefits: How do I count the ways?

First, I can tell you without hesitation that I do not know a single farmer who has converted to grass farming and the production of pasture-raised meats who has switched back to conventional methods of raising animals. Whenever the members of the Maryland Grazers Network get together, someone invariably says, "I can't believe every farmer doesn't do this." Not one of them can imagine going back to confinement-dairy or grain-finishing cows.

Grass farming provides several marvelous quality-of-life benefits for farm families:

- It is a whole lot healthier and more pleasant to be out working in pastures rather than in dusty barns and running heavy machinery.
- Capital costs for equipment and inputs are much lower for new and younger farmers who are just getting started.
- A big reduction in animal health problems lowers costs for farmers, and provides even greater value in terms of a farmer's peace of mind.
- Costs associated with producing pasture-raised meats are lower, and farmers have more control over how they market their animals.
- Grass farming builds healthy soils that enhance a farm's long-term productivity and sustainability. On the other hand, grain cropping with fertilizers and pesticides provides short-term economic returns, but often depletes the farm's long-term productivity.

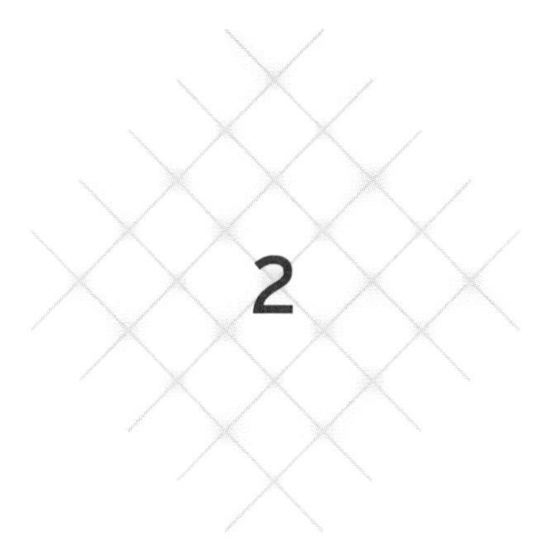

2

Grill Talk

by Rita Calvert

Grassfed, or 100 percent pasture-raised meat, is cooked differently than conventionally raised meat, mainly (but not exclusively) because it is leaner. We all know what happens when we move around a lot: we lose fat. The same is true of animals. Leaner meat also means more pure protein, which means shorter cooking times and, other than searing, lower temperatures on the grill.

Overcooking is the main problem most people have when they switch to pastured meat, so err on the side of less well-done meat (because grassfed meats are leaner, there is a much smaller margin of error). Less is definitely more. You can always throw a cut of meat back on the grill for a minute or two, but you can't uncook a well-done steak.

Another general grilling rule that is especially appropriate for larger cuts of 100 percent pastured meat is *low and slow.* Start with higher heat to sear or caramelize the meat's exterior. Where the fire hits the surface of the food, sugars within the meat melt and yield a beautiful and extremely tasty crust. You are looking for golden brown—not dark brown or charred. You can then finish by moving the meat to the indirect heat area of the grill, often with some full-flavored liquid added.

A SHORT COURSE ON GRILLING, BARBECUING, AND SMOKING

Auguste Escoffier, grandfather of classic cuisine, described grilling as "the remote starting point; the very genesis of our art." It is a passionately debated topic; avid grillers love to compare notes about how they grill.

Grilling

Grilling is best suited for foods that can handle quick cooking, such as steaks and smaller pieces of poultry. Keep in mind that it is a healthful method of cooking as long as you avoid a charred black crust on the food. This rule applies even to vegetables.

It is always important to pay attention during grilling; in other words, keep your eye on the food, even when you are creating a low-and-slow masterpiece, and take an occasional peek if the grill lid is closed. Strike a nice balance between mastering a relaxed calm of experienced grillers and frantic obsession. Practice and you will get to know your grill and gain a greater level of comfort.

Once you become more comfortable with your grassfed grilling skills, you can start to experiment to find your own style and preferences.

Begin by placing meat a few inches from the flame; quickly sear the outside to concentrate the juices inward to retain moisture and to ensure tenderness.

Barbecuing

The technical definition of barbecuing is *cooking with smoke,* but it is often a combination of grilling and smoking. Barbecue also refers to the event or meal in which this type of food is served.

The art of barbecuing is the process of preparing foods that require smoke, low temperatures, no direct contact with a flame, and longer cooking times. Compared with grilling, which demands speed and heat, barbecuing is gentle. Hot smoke does the cooking, gradually adding its own flavor to the meat, which remains naturally moist.

The meats typically chosen for barbecue include pork shoulder, brisket, ribs, mutton roasts, whole hogs, and other beef and pork roasts. It is ideally suited for large pieces of meat such as butts, ribs, or whole smaller animals like pig, lamb, and goat. However, barbecuing is not for sausages (it really belongs in the category of "grilling"). The temperature range of most barbecues—190° to 300°F—is too high for smoking sausages and, as their fat melts away through their casings, they become unappealingly greasy on the outside while the interior can become too dry. This is important to know because sausage, produced from combinations of ground meat, herbs, spices, and seasonings to make use of less popular cuts of meat, have become very common.

Smoking

Smoked food is slow-cooked on a grill away from the fire. Smoking can involve liquids, such as water, beer, wine, or even carbonated beverages that are placed in a container

Ideal temperatures and timing

Grilling: High heat, 400° to 550° F. Hot and fast, mere minutes.
Barbecuing: Low heat, 190° to 300° F. Low and slow, a few hours.
Smoking: Very low and usually indirect heat, 52° to 140° F. 1 hour to 2 weeks, depending on temperature.

---

between the heat source and the food being smoked. Another strategy is to put the meat in a flat, metal, roasting-style pan and place the pan on heavy aluminum foil (you can also put the pan directly on the grate—our recipes include these directions). The hot, moist air helps the meat retain more of its juices and, because the pan is infused with different smoking ingredients, it adds more seasoning to the food.

Avoiding Health Hazards

The true hazard in any burned portion of grilled food is charring, which is when food turns into charcoal from overheating and then burning. Charred meat produces PhIP, a compound shown to cause various types of cancers in rats. A scientist from the University of Maryland warned me that even a charred coating on vegetables is problematic, so stay away from charred food in all cases. Grill marks when the food is seared is fine, you just don't want ash. Carefully remove all charred portions from grilled food prior to serving.

In humans, heterocyclic amines, or HCAs, found in smoked foods pose risks of prostate, colon, stomach, and breast cancers. Another concern arises when polycyclic aromatic hydrocarbons, or PAHs, a compound that forms when fats from chicken, steak, and some types of fish (salmon, orange roughy, and grouper) are redeposited on barbecued meats. In "Grilling for the Novice" (page 30), we suggest you trim all fat to one-quarter of an inch. Now you have another reason since doing so greatly reduces the likelihood of PAHs forming.

RULES-OF-THUMB FOR OUTDOOR COOKING GRASSFED MEATS AND POULTRY

- Allow food to come to room temperature (unless you are in the tropics and without air conditioning).
- Do not precook grassfed products before grilling.

- Preheat the grill at least 10 minutes before grilling.
- Have your grill station prepared and ready with extra table space, a timer, sometimes a small clock, grilling equipment, hot mitts, and a rag.
- Keep a reliable thermometer at your side. There are good thermometers available in a wide range of prices. My $10 digital version has an on-off button that saves the very small battery. On the high end of the scale, you can spend a lot more money for an instant-read digital with a cord and needle that pierces the meat.
- To retain its precious juices, always handle your meat with tongs rather than a fork. Delicious juices will be lost if you slice into or poke the food before you're ready to serve. If you cut into it right away, the juices immediately spill out resulting in a drier texture.
- Once you have removed the food from the flame let it rest, covered loosely with foil for 8 to 10 minutes. This will help redistribute the juices inside the meat before serving.

CHOOSING THE RIGHT GRILL

Grilling, barbecuing, and smoking can involve multiple pieces of equipment. However, all you really need to create exceptional grilled foods are an adjustable grill rack, a lid or cover, an open fire, a few no-frills tools, and high-quality meats. A variety of grills speak to a particular culture or style of cooking, such as the *parrilla* of Argentina or the Tuscan grill of Italy. However, most of the equipment available for outdoor cooking is either superfluous or an unnecessary indulgence. It seems that grilling has become extremely complicated in our culture. Not so in this book. We prefer to keep the message and the techniques simple. While I am not one to be impressed by designer labels, as a culinary professional I find that some brand names speak to quality. However, whenever possible, I like to keep it low-key and save the money for quality ingredients.

Here is a brief overview of the various styles of grills. You can investigate on your own and find the brands that suit your personal style and budget.

Gas and Electric Grills

Because grilled foods tend to cook quickly, gas and electric grills are the easiest to use. Gas grills produce nice results and offer the home cook considerable flexibility. Moreover, gas grills provide consistent heat, which is important when you want to slow down the cooking process. They can burn hot enough to make distinct sear marks on meat and to flavor the finished product with the smoke produced from drippings that fall to the bottom of a hot grill. Today's newer electric grills are almost as hot as their gas counterparts and they have become much more user-friendly, but my first choice is a gas grill infused with a bit of smoke creating fruitwood chips for authentic wood-fired flavor (see "Grilling for the Novice" on page 30).

Radiant Grills

I recently discovered this kind of grill. After trying out my friend's new purchase, I decided to investigate further so I could understand how it works. Radiant (or infrared) grills use ceramic tiles that emit radiant heat and can quickly rise to a steak-searing 900+°F, which is great for cooking a slab of meat to a crunchy crust and for cutting your cooking time, said David Kamen, a professor at the Culinary Institute of America and grill evaluator, in an issue of *Popular Mechanics*. But watch out: high heat makes it easier to burn your steaks or, in other words, get a little out of control. So this grill seems like equipment I would prefer to leave for culinary schools or restaurants.

Wood-Burning Stone or Brick Grills

Here you have the royalty of backyard grilling. These grills are usually made of concrete or brick and lined with special firebrick (a block of refractory ceramic often used to line furnaces and kilns). One unique benefit of these grills is the ability to adjust the height of the racks. The racks are usually one or two pieces with a network of chains that allow the grill master to use hand cranks to raise or lower the height. For the epitome of style, the racks have V-shaped strips of metal that allow the grease to slide down into a tray at the bottom, which helps avoid flare-ups.

Then there is the semi-permanent brick or stone grill, the type of thing that one ad libs by piling up bricks or stones and topping with a grill rack in order to cook real foods with real wood. The masonry wood-burning grill and oven is quite a masterpiece if you're willing to make the investment. You can do it yourself or hire a professional to handle the construction. This is a permanent structure that is fitting for the serious grill aficionado. You can search the Internet to see pictures and get more information about these types of grills. One in particular is the Firepit and Grilling Guru (www.firepit-and-grilling-guru.com/outdoor-stone-fire-pit.html) and the other is Asado Argentina (www.asadoargentina.com).

Flat-Rack Grills: The Parrilla

With grilling becoming an art form, we now hear much more about grilling styles from other cultures.

According to Francis Mallmann, author of *Seven Fires: Grilling the Argentine Way* (Artisan, 2009), Argentineans live their lives around fire. In his home country, the preferred grill is the *parrilla*. Otherwise, simple flat-rack grills, either prefabricated or homemade, are quite popular for their simplicity and low cost. All you need is a surface on which to grill the meats and something to support the surface. In South America, it is common to see construction workers cooking up some chorizos for lunch on top of welded rebar. The one drawback of the flat-rack grills or parrillas is that nothing protects the grill from the weather while cooking. The Tuscan grill has been around in Italy since the 1800s and it is just beginning to travel to the United States. With its simple, heavy, cast-iron, adjustable-rack system, the grill can be adapted to fit inside an indoor fireplace or to work outside over a fireplace or hearth. Alice Waters has been cooking with one for more than 20 years and claims it makes grilling a pleasure (www .tinyurl.com/498f3gp).

Hibachi

In Japan, *hibachi* means *fire bowl.* It originated in China and became popular in Japan as a simple heating device (often bowl- or box-shaped) containing charcoal. In the West, hibachis became popular as small cooking stoves or iron grills that use charcoal. They have since become indispensable as transportable, open-fire cooking equipment. A hibachi is usually the first grill people can afford for those beach bashes or after-game events.

Kettle Grills

Kettle grills create effective air circulation. On the downside, the rack height is not usually adjustable so it is necessary to remove the hot, food-laden rack to add more fuel.

Fire Pits

In *The Joy of Cooking*, Irma von Starkloff Rombauer and Marion Rombauer Becker state that pit cooking "is the most glamorous of all primitive types because it is so largely associated with picturesque places, hearty group effort, and holiday spirit." Fire pit cooking is one of the oldest methods and it is still popular due to its versatility: dig a hole in the ground, fill it with fire, add a large animal, cover, and cook. Most people recognize it as the New England clam bake, the Hawaiian luau or, more specifically, the kalua pig.

In its most basic form, a fire pit is a hole dug into the ground and lined with heavy rock to retain the heat. The food is lowered into the pit, covered with a layer of vegetation, such as banana or taro leaves, and then topped with dirt. The moisture of the earth combined with the fire's heat is what cooks the food.

Manufactured portable fire pits are available today. They often are described as a fire pit vessel and they typically are made of metal, cement, or stone.

Whichever style of grill you choose, be sure to select one with an adjustable rack that can be moved toward or away from the flame and that has a lid or can accommodate an improvised cover. It is always convenient to have a grill that can easily be moved with wheels that lock. High winds and rain may require you to reposition the grill. Since I grill year-round, I like to have options for overhead protection from storms, even if it is the eave of the house. (Important: My house is stucco, not wood!) And be sure that the grill rack has air circulation beneath and around it, as described next.

THE ROLE OF AIR CIRCULATION

Controlling the air and its circulation means controlling the fire. Grill vents are necessary for airflow inside a covered grill for those times when the grill is completely covered. Conversely, if you are grilling in a wide-open space, such as an unobstructed outdoor area, you will need to protect the fire from excessive wind. Whether it is a temporary wind barrier or a wall, some kind of shield is a must.

When cooking in a fire pit, you will have success only if you conquer air movement, if for no other reason than to ignite the fire. When I am giving instructions on building a rudimentary fire, I like to use the teepee design, which gives air circulation throughout. One Mennonite farmer who gave us his recipe for cooking in a fire pit suggested starting the fire by simply laying 3-inch-diameter dry saplings over the opening in the ground and lighting them. After the saplings have burned to embers, they fall into the hole and there you have the fire in the pit. These small tree trunks can only start the fire if there is enough air to fuel the flame when they are propped on top of the pit. You will see our star example of how to cook with a fire pit in the Turkey-in-the-Hole recipe (page 132). You can use this technique for a fire pit for roasting almost any large cut of meat, whole poultry, or even a whole pig as in a luau.

OUTDOOR COOKING FUELS

Natural Hardwood

I have tried just about every kind of (safe) fuel imaginable for grilling and smoking foods. The range spans from corncobs to nutshells. Fruitwoods lend a fine character to grilled foods but I, along with a number of professional chefs, make apple wood my first choice. My friend chef Mark Salter of the Inn at Perry Cabin in St. Michaels, Maryland, said that if I ever decide to cut down my old tree, he would love to buy enough to fill his pick-up truck. (Maybe I should suggest the wood of the fig tree, which I have used for an even richer smoky flavor.) Cherry wood also adds a sweet note, but the wood is expensive because it is used frequently for building quality furniture. Hardwoods, like oak, fruitwoods, hickory, and mesquite give the best results and they give off the hottest flame. Avoid resinous woods, such as pine, cedar, or members of the evergreen family that produce thick, black, turpentine-like smoke.

Natural Hardwood Charcoal

I prefer the real stuff (fruitwood or hardwood) for a live fire because it imparts a clean, clear taste and burns at such a high temperature that browning comes easily. They often can be found in supermarkets along with natural hardwood charcoals. The natural lump charcoals or briquettes are wood that has been previously burned. They create less heat and smoke because the briquettes have expended at least 50 percent of the heat, leaving mostly carbon to use as fuel.

I like combining wood with other heat sources to give foods a smoky taste even when grilling over a gas fire. Try soaking a few pieces of fruitwood in water, placing them on a sheet of heavy foil punctured with a few holes, and laying this open package on a gas, electric, or charcoal fire; the wet wood will smoke for a long time. If you do not have a wood source, numerous varieties of wood chips are available at supermarkets, garden centers, and most outlets that sell grilling supplies.

Charcoal Briquettes

Charcoal briquettes are acceptable for quick starts. I suggest using a chimney starter to get the flame going quickly. Charcoal is a commonly used source for generating heat and is a popular choice for picnics and camping. It burns hot and it is excellent for grilling and easy to transport. This kind of fuel has many devotees but, in my opinion, wood fuel adds much more interest to grilled food.

Never use chemical-based, instant-light charcoal briquettes or lighter fluid. The chemicals not only overwhelm the taste of the food but they also may leach into the food and create health hazards. At my former café in California, I used charred mesquite wood chunks sold in 50-pound sacks. This natural chunk charcoal burns very hot, but for a shorter period of time than the more commonly used briquettes.

Propane

Propane can be controlled easily and it is considered the safest to use for outdoor cooking. The chances of flare-ups are reduced with propane because you can quickly turn off the gas and close the lid. Propane tanks are also efficient because they can be turned on and off, but you do have to worry about keeping them full; a small pressure gauge is helpful. Most gas barbecue grills ignite with the touch of a button so you can forget about the matches and lighter fluid.

Preparation

It is always best to give your grill a good cleaning if you have not used it for a while: turn the heat to high and close the lid for about 10 minutes, then clean the grate with a grill brush to remove the caked-on oil and grease from past grilling. Also clean the grill after cooking to prevent build-up of grease and the remains of charred foods.

INGREDIENTS AND FLAVORINGS

If you chat with your farmers about their pastured meats, they will most likely encourage you to try their meats without salt and pepper so you know how the cuts taste before flavoring them. I agree. Tasting the meat without flavor enhancements will help you understand the special nature of the product. However, once you become more familiar with the flavors, you should be inspired to take your recipes a bit further. While high-quality ingredients matter, there is no need to import truffles from Italy or invest in that $59 bottle of sherry vinegar. Instead, think about including locally grown, seasonal produce and fresh herbs. Back in the old days of my cooking career, I witnessed a farmer or two taking the greatest care with their artisan farm products only to incorporate a mass-produced sauce or seasoning. These days, with high-quality, locally grown and raised ingredients being so much more readily available, there is no need to resort to packaged sauces, spices, marinades, and rubs if you can avoid them. Keeping your food close to the edge of pure flavors can only enhance your cooking ventures.

Certainly salt and pepper have varying levels of quality. Salt is one of the most effective (and most commonly used) of all food seasonings and preservatives. Natural sea salt or kosher salt are highly recommended over common, mass-mined, refined table salts, which almost always have additives, such as potassium-iodide and aluminum silicate to prevent caking. Sea salt basically means unrefined salt that is directly derived from the sea or ocean. In contrast to table salt, natural sea salts retain traces of various minerals such as iodine, zinc, manganese, potassium, calcium, magnesium, and iron. Peppercorns are recommended whole and ground in the moment with a sturdy, adjustable pepper mill. In particular, I recommend Tellicherry peppercorns because they lend fruitiness as well as heat.

Spices and herbs are remarkable ingredients for enhancing food. Spices, being the dried seed, fruit, bark, or root of a vegetable substance, will remain fresh longer if kept in their whole form. Dry roasting and then grinding these spices is a technique common in India and the Middle East that imparts a dimension of flavor and often kills harmful bacteria. Many of our recipes call for this technique and we have been impressed with the flavor of dry-roasted coriander, cumin seed, fennel, or even peppercorns. Herbs are the leaves or flowers of the vegetable plant. The most pronounced flavor comes from using herbs when they are fresh and just picked.

There are times, however, when dried garlic, bay leaf, and onion powder are indispensable. That is when a multi-ingredient spice blend, such as a dry rub or even the "wet paste" for our Java-Pasted Brisket (page 64), comes in handy. Once you discover your favorites, these spice blends can be made in large amounts and stored in sealed jars for frequent use.

BRINING AND MARINATING

When grilling, barbecuing, and smoking pork, lamb, or poultry, consider brining, marinating, or rubbing the meat with a spice blend beforehand for added tenderness and, of course, flavor. I personally have a great fondness for brining, which probably stems from my grandparents' influence; their brine removed all gaminess from that holiday turkey. Their brine also helped retain moistness and tenderness. The simplest brine is a mixture of 1 cup of salt and 1 gallon of very cold or ice water. Submerge your meats up to 24 hours before grilling. The bird should be completely submerged in the brine. If not, it needs to be turned often so its surfaces meet the brine. For crispy poultry skin, make sure to pat your chicken or turkey dry after removing it from the brine and refrigerate for a couple of hours before cooking.

Marinades have three main components: acids, aromatics, and fats. The combination of the three can be mellow and generic or they can lend a distinct ethnic profile.

When our recipe calls for

salt	We mean kosher
pepper	We mean freshly ground and, preferably, Tellicherry peppercorns
olive oil	We mean extra virgin
eggs	We mean free range and large
cheese	We mean artisanal cheeses made with care in small batches
parsley	We mean Italian flat-leaf parsley
wine	We do not mean cooking wine; we mean the real thing

As we mentioned earlier, acids act as a tenderizer and allow the spices, herbs, and other seasonings in the marinade to flavor the meat. Oil acts as a transport vehicle that moves the taste of the aromatics, herbs, and spices into the meat or poultry that is being marinated.

Marinades should be mixed in nonreactive containers, such as stainless steel, porcelain, clay, or food-safe plastic—cast iron and aluminum are unacceptable choices for this purpose.

Acids include all types of vinegars, wines, fruit juices, coffee, and cultured milk products such as yogurt. When used sparingly, flavored vinegars, such as balsamic from Italy, add sweet notes of flavor. A robust apple cider vinegar, such as Braggs, adds a rich flavor to a pork marinade. Wine and wine vinegar are commonly used in European-style marinades, while rice vinegar is commonly used in Asia as part of simple fish marinades. Citrus and some other types of fruit juices shine in marinades; the souring agents of lemon juice or pomegranate juice are often found in Middle Eastern cuisine whereas lime juice is common to Latin American cuisine. Meyer lemon, orange, tangerine, or grapefruit juices can be used for variation. Dairy-based marinades include the yogurt-and-spice mixtures for lamb found in the Middle East; yogurt, cinnamon, and cayenne for India's tandoor; and buttermilk in the American south.

Fats, namely oils, in marinades seal in flavor and help keep foods moist during grilling. Olive oil or oils with mono- and diglycerides penetrate deeper and faster than other types of oils. As with other ingredients, oil can lend an ethnic profile to a dish. Olive oil is preferred in both the Mediterranean and the United States. Flavored nut oils, such as walnut, hazelnut, or sesame, provide a balance to the acids and aromatics.

Yogurt, which has an acid component, also provides fat and it is the foundation of some of the easier-to-prepare marinades. More complex marinades balance acids, aromatics, and fats. A heavy, fruity olive oil needs a rich wine or balsamic vinegar for balance. The cook should avoid using, or only lightly use, strongly flavored oils, such as Asian toasted sesame oil.

AROMATICS

Aromatics are any of a variety of plants, herbs, and spices (such as bay leaf, ginger, and parsley) that impart a lively fragrance and flavor to food and drink. Aromatics add a distinctive character with spicy hot, sour, or sweet flavors.

For all of these flavorings, we do recommend fresh herbs when possible; however, spices (meaning dried) and a few select dried herbs infuse food with remarkable character. These include allspice, bay leaf, juniper berries, mustard seeds, oregano, and peppercorns. Strongly flavored condiments, such as Tabasco, Dijon or spicy mustard, soy sauce, fish sauces, or Worcestershire sauce, add intense bursts of flavor to marinades. Chiles—including ground chile powder, or the smoky ancho adobo chile—are the foundation for recipes of the American Southwest and Thai cuisine and are used in many Latin marinades. Be forewarned: The wide variety of chiles offers many differing degrees of heat.

NATURAL TENDERIZING AGENTS

Tenderizing breaks down collagen in meat in order to make it more palatable. There are numerous ways to do this but many of the more commonly used products are chemically based or can cause adverse reactions. Here we offer a look at some of the more natural tenderizing ingredients, all of which are extremely effective.

Acid marinades and the plant enzymes found in papaya, figs, pineapple, and fresh ginger break down the muscle fiber and collagen in meat and poultry that lead to some tenderizing. Bathing meat or poultry in acid begins the cooking process. In *On Food and Cooking: The Science and Lore of the Kitchen* (Scribner's, 1984), Harold McGee describes how enzymes have been used for hundreds of years. According to McGee, in pre-Columbian Mexico, meat was wrapped in papaya leaves prior to cooking. Yogurt and buttermilk also contain calcium, which activates enzymes in meat that break down the muscle fiber. It is conceivable that the calcium in yogurt may result in the same sort of tenderizing that you get when you age meat.

You may want to consider these natural tenderizers.

- Acids, such as apple, tomato, and citrus juices for marinating
- Buttermilk
- Coffee, strong dark roast brewed and cold
- Figs

International flavors

If you are interested in achieving a basic ethnic flare, there are a few ingredients we recommend, some of which may travel from distant shores. Chopped ginger will suggest an Asian influence in a teriyaki marinade, especially if used along with lemongrass, soy sauce, and garlic. Chinese-style marinades typically use ginger, scallion, garlic, and possibly Hoisin or black bean sauce. A mirepoix—finely chopped onions, carrots, celery, and often leeks—is what flavors French-style marinades. Tumeric, coriander seed, cumin, mustard seed, and cardamom are most frequently used for Indian-style cuisine. Latin-style marinades also feature lots of garlic, cumin, chiles, and lime juice.

- Ginger root (fresh)
- Kiwi
- Papaya
- Pineapple
- Salt and water for brining
- Tea, strong brewed and cold
- Vinegars such as balsamic, rice wine, apple cider, red and white wine, pomegranate
- Yogurt

These ingredients are often blended with spices for the purpose of flavoring. It is best to start with a recipe that incorporates a key natural tenderizing agent. We use many of these in this book. Once you are familiar with the effect that particular ingredient has on food you can create as you like.

MAKE IT SEASONAL AND FRESH

Seasons form a natural backdrop for eating what is meant to be eaten during the year. Pastured spring lamb all but cries out for fresh asparagus followed by a dessert featuring the first blush of strawberries. Depending on where you live, summer's grilled bounty of steaks, burgers, and barbecued chicken marry harmoniously with a salad of heirloom tomatoes and fresh basil along with a cob or two of bread-and-butter corn. The holiday bird announces fall, complete with apples, pears, delicata and butternut squashes, sugary sweet from their sojourn in the hot summer sun along with the abundance of rainbow purple, blue, orange, and golden potatoes.

A word about garlic

You will notice in the recipes that when preparing sauces, stews, or soups, I add garlic toward the end of the cooking process. My theory is that the essence of the garlic gets lost when sautéed early in the process and then gets even more lost when the recipes require simmering down, possibly for hours. When I shared these thoughts with another cook, he agreed but explained that he wanted the fragrance of garlic to fill the kitchen from the start, so we compromised: sauté garlic at the beginning, cook it down, and then add more toward the grand finale. You might use twice as much, but that's a good thing! Notice that we give the size of garlic cloves in the recipes. The size of a clove can dramatically change the recipe. Elephant garlic is not used in this book–just your basic garden-variety garlic, although clove sizes vary from huge on the outside to very tiny near the center.

The broad view of eating seasonally is to live in harmony with nature, to connect with our core values, and to support our local farmers, our local economies, and our local systems of agriculture. Eating fresh and seasonally is not a new concept; it was, in fact, the only choice for many societies before the economy went global. Today we eat locally and seasonally as a way of reclaiming our food roots and doing so benefits our health, our communities, and the environment.

Remember that seasonal foods are foods that are produced in your region; they are, therefore, local by definition. Foods described as *in season* are at their peak of freshness, and often they are a better value.

DEFROSTING GRASSFED MEAT AND POULTRY

Your farmers have invested enormous amounts of time and effort into raising the 100 percent pastured meat you have purchased. You may have invested some time getting to know nearby farmers and understanding their products. You may have spent a good deal of money on the meats you purchased. Make sure you take the utmost care in how you store, handle, and prepare them.

Packaging for meat and poultry has changed dramatically in the past few years. The meat you buy may be packaged in butcher paper or it may be sealed in plastic. The "vacuum bagging system" (or Cryovac system) is preferable to butcher paper because frozen products can last up to 18 months in the freezer with if they are sealed in heavy-weight plastic. Freezer life is determined by the grade of plastic used but grade of plastic affects the price of the product—the stronger the plastic, the higher the cost to the farmer, and that usually gets passed along to the consumer. Again, the best way to understand the entire process is to discuss meat and poultry packaging with your farmer.

If you are purchasing your meat frozen, I recommend two safe and reliable methods for defrosting: refrigerator defrosting and cold-water defrosting.

Refrigerator Defrosting

When refrigerator defrosting, planning ahead is the key because of the length of time involved. A large frozen item—such as a turkey—requires at least 24 hours for every 5 pounds of weight. Even small amounts of frozen food, such as a pound of ground meat or boneless chicken breasts, require a full day to thaw.

When thawing foods in the refrigerator, consider these variables:

- Some areas of the refrigerator may keep the food colder than other areas. Food placed in the coldest part requires longer defrosting time.
- After thawing in the refrigerator, ground meat and poultry should remain usable for an additional day or two before cooking; three to five days for red meat.
- If your meat or poultry is wrapped in butcher paper, refrigerator thawing is the only method to use. Make sure you put the package on a plate or bowl to catch leakage.

Cold-water Defrosting

Cold-water defrosting is faster than refrigerator thawing but it requires more attention. The food must be in a vacuum-sealed, leak-proof package, or plastic bag. If the bag leaks, bacteria from the air or surrounding environment could contaminate the food. Also, meat tissue can absorb water like a sponge, which will mean a watery product.

Submerge the sealed bag in cold tap water and change the water every 30 minutes while the food continues to thaw. Small packages of meat or poultry—about a pound—may defrost in an hour or less. A 3- to 4-pound package may take 2 to 3 hours. For whole turkeys, estimate about 30 minutes per pound.

Cold-water defrosted food must be cooked immediately: freeze only after cooking.

Never ever defrost meat by submerging it in warm or hot water.

Avoid defrosting meat by leaving it on a countertop at room temperature. Doing so invites bacteria that can contaminate the meat. Defrosting in the fridge keeps meat cold enough to ward off bacteria.

We strongly disapprove of microwave defrosting, mainly because foods in a microwave thaw inconsistently. You invariably will end up with edges that are cooked while other sections remain frozen. This is an invitation to bacteria.

GRILLING FOR THE NOVICE: A PRIMER

by Rita Calvert

I want to share this confession from a dear friend who loves farm-fresh food and vows to spend as much time outdoors as possible.

"As much as I adore grilled food, I do not know how to do this myself. I don't even know how to turn on the gas grill or build a charcoal barbecue fire."

She's game to try it now with the same easy tips I am sharing here with you. You can find more details in many other books and websites but this basic primer will get you started and reduce the stress of this oh-so-versatile cooking method that makes food so tasty. With this primer and the recipes you'll find in *The Grassfed Gourmet Fires It Up!*, you will be barbecuing all summer (or year-round) like an old hand. Start by practicing on yourself or with a small group.

Plan ahead

You can cook just about anything on the grill. With just a little planning, you will have your indoor kitchen clutter-free and cleaned up, even before serving the meal, regardless of whether you're cooking for yourself, your family, or a whole gang of friends. Best of all, you won't be washing loads of pots and pans in the wee hours after a relaxed meal. If you plan side dishes that are safe to serve at room temperature, such as breads and simple vegetable and potato side dishes, you will have even more freedom to prepare ahead and then relax.

For a crowd, you can prepare side dishes ahead (or assign sides to your guests), present them on serving platters, then completely clean up your indoor kitchen before the meal.

If you're cooking for a party, get the basics—such as the centerpiece, plates, flatwear, serving pieces, and some condiments—on the table before guests arrive. This will save time later in the evening when your food is ready for serving.

Set the stage for grilling

Before you fire up the grill, be practical. Position your grill away from all shrubbery, overhangs, grass, or other flammable surfaces or materials. Don't even think about a closed area like a garage. Your grill should be on a stable surface. Never use gasoline or kerosene to start your fire. And make sure the grill is clean and the vents are open and airy—they should not be clogged with old ashes.

Choose your grill

When it comes to quick and easy cooking, a gas grill is probably the most convenient way to go, but nothing beats the flavor of foods cooked over real wood or charcoal. Not to worry, though, because you can do both at once. Add real wood flavor to your gas-grilled foods by placing small green or water-soaked wood chips on foil that has been punctured with a few holes to let air circulate. Start the grill and then place the package on a portion of the gas element and let it start to smoke. If you are using a charcoal grill, be aware of the time it takes to prepare the fire and heat the grill before you start cooking. This can vary depending on the type of wood or charcoal you use, so test different brands beforehand and stick with the one that works best with your grill.

Prepare your grill

Prepare the grill by rubbing clean grates with cooking oil using a paper towel. Food won't stick and cleaning up is far less messy and stressful.

Always preheat your grill—with the lid closed—for the time recommended for the brand and style of your grill or to the suggested temperature in the recipe you're following. Gas grills need to be turned on at least 10 minutes before cooking. If you use a charcoal grill, allow the coals to burn for at least 30 minutes or until the flames subside before cooking.

Note: Always keep the bottom tray and drip pan of your gas grill clean and free of debris. This not only prevents dangerous grease fires, but it also deters visits from unwanted critters. A sprinkle of red pepper in the pan is another safe way to discourage animals.

Light your grill

For gas grilling:

Read all grilling instructions that came with your grill first; every grill ignites differently. In general, however, you start by opening the grill lid, then opening the tank valve, then turning the front/first burner to high heat. Allow 2 to 3 seconds for the gas chamber to fill. Then push the igniter button firmly. The burner should light after only one or two pushes of the button. Once the first burner is lit, turn the middle/next burner to high heat and repeat with the other burners until all burners are lit. Close the lid. Allow the grill to preheat on high to 500° to 550°F. Place your food on the cooking rack and adjust burners to the temperatures and cooking method given in the recipe.

Consult your grill's instructions about what to do if there are flare-ups. In my experience, when flare-up occurs, I turn all burners to the off position and then move the food to another area of the cooking grate. Then, I light the grill again. Never use water to extinguish flames on a gas grill.

For charcoal grilling:
I have found the best method for firing up a charcoal grill is with a quick-start chimney. They usually sell for about $15 where grills are sold. Fill the bottom of the chimney with some crumbled newspaper, then place charcoal or briquettes on top of that, and ignite. In about 25 minutes, your coals should be ready. The charcoal will be lightly coated with ash. Carefully pour the heated charcoals out of the starter and into the grill. Arrange them evenly across the charcoal rack for the "direct method" of grilling or on either side of grate for the "indirect method" of grilling. New terms? See below.

Another option for starting your charcoal grill is to place crumpled newspaper or fuel cubes on the charcoal grate. Cover the paper with charcoal briquettes to form a pyramid (not too huge) and then light the charcoal. It should be ready in about 25 minutes when a light grey ash coats all of the briquettes. We recommend learning to do this without using lighter fluid to keep your grilling simple, clean, and safe.

Choose your grilling method:
Before preparing your grill, decide if the food you are grilling requires "direct" or "indirect" heat. With a little practice, these two approaches to cooking will become second nature—and you will have mastered one of the most important grilling techniques. Remember, though, whether you are cooking with the direct method or indirect method, always grill with the lid on.

Grilling with direct heat is similar to broiling except that the heat source is below the food; in other words, the food is cooked directly over the heat source (thus "direct" heat cooking). Use the direct method for foods that take less than 25 minutes to cook, such as steaks, chops, kebabs, sausages, and vegetables. Direct cooking is also necessary in order to sear meats.

The indirect method is similar to roasting but with the added benefit of grilled texture, flavor, and appearance, which you can't get from an oven. This is best for foods that require 25 minutes or more of grilling time or foods that are so delicate that direct exposure to the heat source would dry them out or scorch them. Use the indirect method for roasts, ribs, whole chickens, turkeys, and other large cuts of meat as well as for delicate fish fillets.

Direct-heat cooking: how-to
To grill on a charcoal grill using a direct-heat method, spread prepared coals evenly across the charcoal grate. Set the cooking grate over the coals and light the coals. Once the coals are ready, place the food on the cooking grate. Close the lid, lifting it only to turn food or to test for doneness at the end of the recommended cooking time.

To grill on a gas grill using a direct-heat method, preheat the grill with all burners on high. Place the food on the cooking grate and then adjust all burners to the temperature noted in the recipe. Close the lid of the grill and lift only to turn food or to test for doneness toward the end of the recommended cooking time.

For even cooking, food should be turned once halfway through the grilling time. Searing creates that crisp, caramelized texture where the food hits the grate. Those nice grill marks add more than visual appeal, they flavor the entire food surface. Steaks, chops, chicken pieces, and larger cuts of meat all benefit from searing.

Indirect-heat method: how-to

To grill on a charcoal grill using the indirect-heat method, arrange hot coals evenly on either side (or around the perimeter) of the charcoal grate. A drip pan placed in the center of the charcoal grate between the coals is useful to collect drippings that can be used for gravies and sauces. It also helps prevent flare-ups when cooking fattier foods, such as chicken or turkey with the skin on, goose, duck, or certain roasts. For longer cooking times, add water to the drip pan to keep drippings from burning. Place the cooking grate over the coals, light the charcoal, and once the grill is heated, place the food on the cooking grate, centered over the drip pan or charcoal grate. Close the lid and lift only to baste or check for doneness at the end of the suggested cooking time.

If you are using a gas grill, preheat the grill with all burners on high. Then adjust the burners on the sides to the temperature noted in the recipe. Turn off the burner(s) directly below the food. For best results, place roasts, poultry, or large cuts of meat on a roasting rack set inside a disposable heavy foil pan. For longer cooking times, add water to the foil pan to keep drippings from burning.

In some cases, it is best to sear the food first to obtain grill marks and then place the food in a cast iron or an aluminum pan to catch the juices for the rest of the cooking time. Heat rises, reflects off the lid and inside surfaces of the grill, and slowly cooks the food evenly on all sides. The circulating heat works much like a convection oven so there's no need to turn the food.

Prepare meat for grilling

For steaks and chops, trim excess fat, leaving only a scant 1/4-inch of fat, which is sufficient to flavor the meat. For poultry, note that the skin has enough fat to feed the flame, potentially leading to flare ups. Less fat is a virtual guarantee against flare-ups and it makes cleanup easier.

After trimming the meat, marinate according to the recipe or rewrap and chill. About an hour before grilling (but for food safety, no earlier), allow meat to come to room temperature.

Cooking tips

Create two temperature zones in your grill: one warm and one hot. If you use a gas grill, turn one side to high heat and keep the other on low. If you use a wood/charcoal grill, the trick is to push most of the embers toward one side. This will help cook pieces of food more evenly by allowing you to periodically move them from low heat to high.

Hold the sauce. If you're using barbecue sauce or any sauce that contains sugar or fat, wait until about the last 15 minutes before slathering it on (if you've marinated your meat in advance, just blot it with a paper towel before placing it on the grill). Since sugar and oil will cause lots of flames and char the food, plan to reduce the heat a bit after you add the sauce.

Allow more cooking time on cold or windy days or at higher altitudes, and less time in extremely hot weather.

Once you put the food on the grate, allow it to cook a bit before any turning. Your food needs about 10 minutes of cooking time for its surface to cook enough to release from the grill easily without sticking and tearing.

Always use a spatula or tongs when you handle the meat on the grill. Using a fork to pierce meat while it is cooking will cause all the yummy juices to escape, thereby drying out your meat.

Use a meat thermometer and a timer so you know when it's time to take food off the grill. Checking meats for internal temperatures is the best way to determine when food is properly cooked or when done is about to become overdone.

Soon after you finish cooking, use a wire brush to scrape and clean the grates—it's so much easier to clean it up when it is still warm.

3

Getting to Know Your Farmer

by Rita Calvert and Michael Heller

The absolute best and most satisfying way to fully understand and appreciate the quality of the meat you buy is to get to know the farmer who actually produced the meat.

What follows are questions you may want to ask a farmer.

DO YOU USE HORMONES OR ANTIBIOTICS?

None of the grass farmers we know use hormone implants to speed the growth of their animals. Most of them use antibiotics only when an animal is sick, which is rare with animals raised on pasture. In most cases, animals that have received antibiotics are not sold as grassfed meat, but it is a good idea to check with your farmer on hormone and antibiotic use.

WHAT DO YOU DO TO ENHANCE THE QUALITY OF YOUR PASTURE?

If you visit a farm, pasture quality is something that you can partly judge for yourself. A pasture should be subdivided, thus allowing the farmer to move the animals every few days. Doing so helps protect the pasture from *overgrazing*. Therefore, it is a good idea to ask farmers how often they move their cows to fresh pasture. The animals should not be grazing on a large pasture where they will stay for weeks. Optimal grazing periods are two to three days on one pasture. Grazing for up to a week on one pasture is not uncommon, but grazing on a pasture for more than a week greatly reduces the pasture's quality.

The pasture should not be overgrazed. You can spot overgrazed pasture by looking for very short grass with bare spots of brown earth showing in places. If your lawn looked that way, you would not need a lawn mower.

Undergrazing can be just as damaging to the nutritional quality of the pasture as overgrazing, so make sure the pasture doesn't look scruffy. If you see areas of tall, brown-stemmed grasses and weeds, the pasture grasses are likely to be less nutritious. After animals have been moved, a well-managed pasture will occasionally be mowed to cut the plants that the cows missed as well as to keep everything young, green, and nutritious. However, farmers may allow pastures to grow tall so that they can cut them for hay or because they want the plants in the pastures to go to seed (fallow), which helps rejuvenate fields. Both are good practices. The radical approach to grazing, called high-density grazing, appears to have great promise when it comes to improving pastures and will often involve pastures with mature, stemmy vegetation. Check with your farmer. If he says that he is a "mob" grazer or high-density grazer, the brown grasses in fields may represent a step forward in grazing management.

CAN YOU DESCRIBE YOUR ANIMALS' DIET IN THE MONTHS LEADING UP TO PROCESSING?

The quality of forage an animal eats throughout its life is important. This is especially true during the last few months of grazing, before an animal is processed. Therefore, it is useful to ask the farmer what type of pasture or forage the cows have grazed during their final few months of grazing. Ideally, they should be put on high-quality pasture (young, green pastures supplemented with high-quality hay rather than older grasses supplemented with moderate- to poor-quality hay).

HOW OLD ARE YOUR ANIMALS WHEN THEY ARE PROCESSED?

Cows should be between 18 and 24 months old before processing; bison between 22 and 28 months; sheep, optimally between 4 and 6 months, and not more than a year old; goats, less than a year. This is important to keep in mind because as animals get older, their meat tends to get tougher.

HOW LONG IS THE MEAT DRY-AGED AT THE MEAT SHOP?

Dry aging is a simple and natural process of tenderizing that involves hanging sides of meat in a cooler. Ideally, beef hangs for a minimum of two weeks for maximum tenderizing; some believe three weeks is optimal, though research indicates that the benefits of the third week are marginal at best. Meat processors and farmers face measurable costs in hanging the meat for two weeks. For the processor, it takes up space in the cooler. For the farmer, the weight loss due to evaporation can reduce the weight of meat by 15 to 20 percent. That means if the meat hangs for two weeks, the farmer is selling less weight. Of course, higher-quality meat offsets this weight loss. Look for beef and bison cuts that have been dry aged 14 to 21 days and lamb that has been aged 3 to 7 days. Goats, chickens, and hogs do not require any aging.

IS THE PROCESSED MEAT FROM YOUR FARM VACUUM-PACKED?

As discussed earlier, vacuum-packed meat has twice the freezer life compared to meat packed in traditional butcher paper and taped closed.

Once you become more familiar with grassfed meats, other questions may arise, such as the following.

WHAT SHOULD I KNOW ABOUT DIFFERENT BREEDS OF ANIMALS?

A variety of opinions exists on the importance the breed of animal has on the quality of its meat. The consensus among the farmers we have talked to is that breeds are not that important in determining meat quality, though they can make a difference to farmers when it comes to production. For instance, most farmers raising grassfed lamb prefer *hair* sheep breeds—such as Katahdin and Dorper—to *wool* sheep. Why? During hot weather, hair sheep tend to graze more than wool sheep, and that means hair sheep tend to gain weight more quickly on pastures. Additionally, hair sheep have a natural, genetic resistance to parasites that are easily picked up by other breeds when grazing.

With cattle, it is better to have animals that mature to a smaller size and achieve tenderness on grass without excessive marbling in their meat. For this reason, breeds such as Angus, Hereford, and Shorthorn are favored, as are breeds that are less common, such as Devon, Dexter, and American Lowline. There is, however, considerable variation within each breed. Knowing the breed alone does not ensure a quality-grazing animal. Occasionally, but not often enough, you will see cows or sheep that are mixed breeds. While most farmers are partial to a particular breed (just as many people are partial to a breed of dog), crossbreeding is a positive farming practice. Crossbreeding provides hybrid vigor that leads to higher fertility rates, faster growth, and gentler dispositions. As a farmer, I am partial to crossing cattle breeds like Red Angus, Shorthorn, and Devon that do particularly well on pasture. However, even crossbreeding can be overdone, which can lead to greater variation in the quality of the animals. When it comes to meat quality, my own feeling is that what the animals eat (pasture quality!) and how they're treated is far more important than breed.

WHERE CAN I FIND GRASSFED MEATS AND POULTRY?

When possible, we recommend purchasing grassfed meats directly from a farmer. If you are completely new to the world of grassfed, you can start by checking www.eatwild.com and selecting your state. Eatwild.com is the brainchild of Jo Robinson, an investigative journalist and author of *Pasture Perfect* (Vashon Island Press, 2004). The site is easy to navigate and it features lists of pasture-based farmers by state as well as information about the benefits (including the health benefits) of foods from animals raised on pasture. If you find a farmer near where you live or work, call or email and find out if you can buy direct.

If purchasing straight from the farm is not practical or convenient, there are other venues for purchasing grassfed meats. These days, old-fashioned butchers and butcher stores are hard to come by. If you can find one, especially one that carries grassfed, you might enjoy the experience of purchasing your meats there, particularly if the butcher has a relationship with the farmers. The benefit of going to well-trained butchers is that they really know their products. They can speak to the quality of a product and they can offer advice about how to prepare a particular cut of meat.

The demand for locally grown and raised foods has given rise to an increasing number of specialty-food stores that carry grassfed meats and dairy products, some of which are nonprofit organizations with a mission to support local agriculture. A good example is Fair Food in Philadelphia. Situated among the Amish farm stands, specialty coffee shops, bakeries, butchers, and deli stands in the bustling Reading Terminal Market in Center City, Fair Food has become the focal point of the local-food movement in the City of Brotherly Love. Its popularity is largely due to the staff's commitment to knowing each and every farmer-supplier and to familiarizing themselves with every product they sell. The shop features an impressive array of grassfed beef, poultry, pork, lamb, and goat products as well as grassfed dairy items. Organizations and companies like Fair Food are good examples of retail at its best. Most of us are busy, with little time to investigate the sources of our food. Patronizing markets such as Fair Food have the added benefit of knowing that the farms have been researched and the products have been sampled and vetted. And nothing is more gratifying than asking a salesperson to help you decide between ground beef from two different farms, and they can.

HOW MUCH MEAT SHOULD I PURCHASE IN ONE TRIP?

The amount of refrigeration and freezer space you have available will help you determine how much meat to purchase at a time. If you don't have much storage space, buy less or think about purchasing an extra freezer; if you are buying meat in quantity, a good freezer is a good investment. But, before you go out and purchase one, look at the entire picture and do the math: if you and your family regularly enjoy cooking and grilling with meat and poultry, a chest or upright freezer in your basement or garage might be a good investment.

HOW LONG CAN MEAT STAY FROZEN AND STILL MAINTAIN ITS QUALITY?

Frozen meat can be stored 9 to 12 months; ground beef can be stored 3 to 4 months. Most cuts of meat can be stored safely for longer periods, but they lose some quality; long storage periods invite freezer burn, dehydration, and broken packages, all of which can compromise quality.

DO I HAVE THE FREEZER SPACE?

A rule of thumb is one cubic foot of freezer space for each 35 to 40 pounds of cut and wrapped meat; allow slightly more space when the meat packages are oddly shaped. You will also have to allow for the difference between buying beef in bulk and buying poultry in bulk. Storing the parts of a whole chicken requires a lot less storage space than what you will need if you purchase an entire cow.

Buying meat does not have to be complicated. Really, there are only three ways to purchase meat:

1. In bulk, as a whole carcass (the entire animal, including all the meats associated with it); a side (half the animal, including hind or forequarters); or a quarter.
2. As a wholesale or primal cut (loin, round, chuck, rib, brisket, foreshank, short loin, flank, and round or rib). These are basic cuts that are the result of cutting carcasses and sides into smaller portions. For the consumer, however, they are still larger portions.
3. One or a few retail cuts at a time (packaged steak or a whole chicken, such as what you buy from your grocer).

If you are buying your meat directly from a farmer, here are a few tips for ensuring a satisfactory experience.

Check before you go: Before going to a farm, confirm that the farmer carries the grade, weight, and amounts of meat you expect to purchase. Some farmers prefer to sell only in bulk. Others work with local processors to cut and package their meats in much the same way you find them in supermarkets.

Check the labels: Make sure that the meat carries the USDA inspection label. Even packaged meat from a farmer should be labeled to that effect. If not, ask if the meat has been inspected.

Portion size: Purchase only the amount of meat you need. Some consumers prefer a carcass or a side of beef; others require only a quarter or one wholesale cut.

Be aware of your preferences: Be sure that the cut of meat you are buying suits your purposes. Some families prefer roasts, steaks, and pot roasts; others are more interested in ground and stew beef. If you decide to buy a half or quarter of a cow, for example, and you find that you have not used all the cuts in a carcass, let that inform your next purchase.

Some farmers will accommodate special requests or will agree to work with you if you have particular preferences. So, if it is important to you, make sure to discuss the thickness of steaks and the dimensions of other cuts. If you want the bones or organ meat, you might have to make a special request.

Understand pricing: When you think about price, keep in mind that the approximate cost per pound of wrapped meat is determined by the price per pound of weight divided by the percent yield of edible meat.

If you are considering buying meat in bulk—particularly beef—here are a few questions to ask and information you may need before you decide:

DOES BUYING IN BULK MAKE FINANCIAL SENSE?

Some people believe it does. Surely when it comes to grassfed beef, it is more cost-efficient to buy in bulk than it is to buy one piece at a time. Buying in bulk can mean a large payment upfront, but if you already have the freezer space you could save a good deal of money.

How many pounds of meat are there in a carcass, a side, a quarter, or a wholesale cut?

An average beef carcass weighs about 600 pounds. Grassfed animals tend to be slightly smaller. A side usually weighs between 250 and 300 pounds (this refers to the hanging weight or the gross weight by which the animal is sold). This can be important because sometimes farmers charge based on the hanging weight and sometimes they charge on a per-pound basis of finished meat. The average weight losses from cutting and trimming make up about 25 percent of a yield from a grade three carcass. That leaves about 350 to 450 pounds of usable meat cuts from a cow or 175 to 225 pounds from a side.

When buying meat it is helpful to know how much fat is on the carcass because this influences the amount of product in the packages.

The rule for carcass beef is 25 percent waste; that is, 25 percent ground beef and stew meat, 25 percent in steaks, and 25 percent in roasts.

What are the advantages of buying a whole beef carcass or a side?

When you buy a whole carcass or a side, you end up with a variety of high-and low-priced cuts. You will find in your order some of the less popular cuts, such as brisket, short ribs, and shank. Usually locker plants and meat markets convert such cuts into ground meat or stew meat.

Meat processors will often age the meat for 10 days, or the period desired, and will cut according to specifications. Therefore, usually you can specify how the side or wholesale carcass is cut. For a carcass or a side to be a "good buy," the purchaser must use every part of the carcass. If the family does not eat certain sections, then it may be wise to have the unacceptable cuts processed into ground beef. If you use a great deal of ground beef, perhaps you should purchase only the forequarter. Buying a carcass or a side may be less expensive per pound, but remember that 24 to 45 percent is lost from cutting, trimming, and boning depending on the yield grade of the carcass. When comparing alternatives, include the cost of cutting, wrapping, quick-freezing, any interest charges you may be accruing if you are financing your purchase, and any finance charges you may be accruing if you have financed your freezer (as well as the freezer's operating costs).

Always purchase large quantities of meat already frozen: be sure the operator or processor has quick-frozen the meat before you take it home.

FYI: Meat should be frozen at –10°F or lower as quickly as possible.

Are there advantages to buying a quarter?

When buying hindquarters and forequarters, you may find that you can be a little more selective. Hindquarters, which include the wholesale round, loin, and the flank, give more steaks and roasts, but they cost more per pound than a side or carcass. Forequarters, which include the chuck, rib, brisket, plate, and full flank, have more of the less-tender cuts. These cuts yield more of the parts that require pot-roasting, but provide a higher percentage of usable lean meat and cost less than the side or carcass.

Is it more economical to buy retail cuts at a market and freeze them?

Buying retail cuts to freeze at home means buying only the assortment of cuts your family prefers, without have to buy seldom or never-used items. Related to this, buying retail cuts is invariably a smaller cash investment than making a bulk purchase, which means less of a need for an expansive—and expensive—storage area.

4

Beef and Bison

Is that a beef cow or a dairy cow?

The black-and-white spotted cows that you see as you drive along the countryside are dairy, not beef, cows. Beef cows tend to be all black, all red, red and white, or black or red with white faces; there are other patterns, but this covers the majority and corrects a common error.

BEEF: FLAVOR IS THE KEY

When you set out to purchase beef, you may have some preconceptions that you will need to toss aside. For instance, when it comes to well-flavored beef, marbling isn't everything. In fact, marbling can get in the way of flavor or, even worse, it can disguise the lack of flavor in meat that comes from grain-fed cows. And yet, USDA standards for beef still use marbling as the main measuring stick for quality: *Prime*, highest marbling and priciest; *Choice*, less marbling than prime; and *Select,* least marbling and least expensive. Well, jettison those old ideas as you begin your adventure with grassfed beef. Flavor is your measure.

Another thing to know about marbling is that the marbling and fat in grassfed beef may have a yellowish color to it. This is a *good* thing because the color comes from pigments, known as carotenoids, in the plants that the cows eat (10). Studies show that when carotenoids figure into the human diet, people are not only healthier in general, they exhibit lower mortality rates from numerous chronic diseases (17).

Grassfed beef can be flavorful and tender, or flavorful and tough—meaning that it almost always has good flavor but often it is not very tender. Most farmers are relearning how to grow pasture-raised beef that is truly tender.

To maximize tenderness, animals should be no older than 18 to 24 months when sent for processing and the meat should hang for 14 to 21 days, *dry aged*, at the meat shop before it is cut into steaks, roasts, and ground beef.

Remember: When it comes to favorite cuts for grilling, costly steaks like tenderloin and filet mignon aren't necessarily the best. We believe that you can't beat the far less-expensive flank, skirt, and hanger steaks for flavor. However, these cuts are easier to ruin. You have to really pay attention during preparation: when grilling, you can go from perfectly cooked beef to overcooked in less than a minute.

BISON OR BUFFALO?

Tradition would have us call these animals *buffalo* but, more accurately, the American buffalo is called *bison*. They are more closely related to cattle than they are to the true buffalos—the Asian water buffalo and the African buffalo. You may have heard of *beefalo*. This is a hybrid, a cross between beef cattle and buffalo. The fertility of these hybrids speaks to their close kinship. In fact, their ability to crossbreed with beef cows threatens the long-term genetic purity of the bison species. Bison once ranged throughout the central grasslands from Canada south to Mexico and as far east as the western edge of the Appalachian Mountains. In the late 1880s, hunting pressure from both the Plains Indians and European settlers moving westward almost drove the bison to extinction. However, bison have made a comeback thanks to a few foresighted individuals who made great efforts to save them by raising and breeding genetically pure strains of bison on ranches.

Bill Edwards, a bison grass farmer, says that bison tend to appear bigger than they are. A full-grown female bison will be only about the size of most beef cows and smaller than most dairy cows. Bill also claims that they act bigger than they are, too. Bison are unpredictable. You shouldn't just wander through a herd of bison on pasture as you might do with many pasture-raised beef or dairy cows. Although farm-raised bison occasionally get agitated, they become acclimated to the farm setting and are less erratic and aggressive than the wild bison you might encounter in Yellowstone National Park.

They are also great grazers, but they require more space for moving around than do beef cattle. This means that pastures must be large enough for bison to roam a bit, which alters the grazing strategy of the bison grass farmer.

Bison are usually 22 to 28 months old when sent for processing. The meat should hang, dry aged, 14 to 21 days for added tenderness. Even though bison meat, which is darker than beef, has no marbling and is quite lean, it is still as tender as beef. Bison, like beef cattle, are ruminants and their meat provides all of the health benefits of their grassfed relatives.

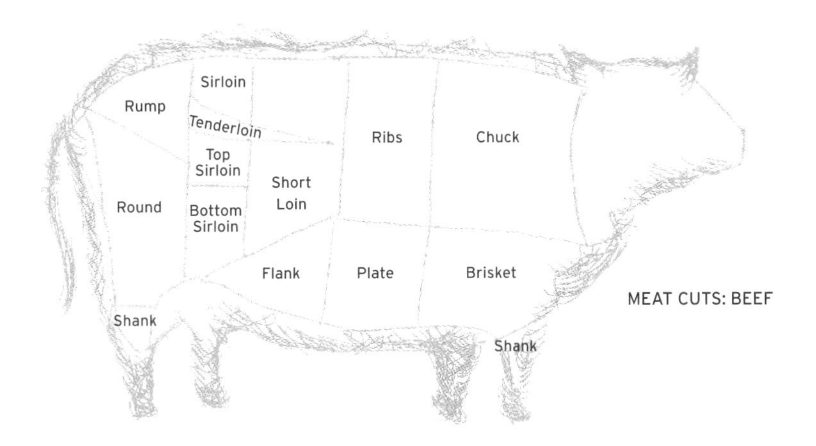

MEAT CUTS: BEEF

Garlic and Herb-Grilled Beef Tenderloin

Let's begin with magnificent and simply prepared beef tenderloin, a premium cut of beef. Choose a whole tenderloin that is on the small side because it will be more tender. There is another, less-expensive but equally exceptional cut: the beef tenderloin butt roast. This portion of the tenderloin is incredibly tender and flavorful. Use this recipe as directed—just make sure to keep it on the rare side.

serves 12 to 14

One 4½-pound trimmed beef tenderloin
⅓ cup extra-virgin olive oil
6 medium garlic cloves, thinly sliced
2 tablespoons coarsely cracked peppercorns
1 tablespoon chopped fresh thyme
2 teaspoons chopped fresh marjoram
2 teaspoons chopped fresh rosemary
2 teaspoons kosher salt

Preheat the grill to medium-high.

Fold the thin end of the tenderloin under the roast to make the meat an even thickness. Tie the roast at 1-inch intervals and transfer to a large, rimmed baking sheet.

In a small bowl, combine the olive oil, garlic, peppercorns, thyme, marjoram, rosemary, and salt. Rub the herb oil all over the roast and refrigerate 2 to 4 hours. Bring to room temperature before grilling.

Grill the roast directly over moderately high heat, turning often, until nicely browned, about 30 minutes for medium-rare or 125°F on a meat thermometer. Transfer to a carving board; rest 10 minutes, loosely tented under foil.

Carve the tenderloin roast into ½-inch-thick slices and serve.

Simply Sublime Bison Steaks

This recipe, which is from one of our farmers, is great for sirloin, round, and flank steaks; it also adds flavor to filet mignon, New York strip, and rib eye cuts that do not require marinating. The ingredients are difficult to identify from the juicy results, but they are simple. Remember to save those precious juices by using tongs to turn the steaks rather than piercing them with a fork.

serves 4

1½ pounds bison steak
1 tablespoon minced onion
2 tablespoons brown sugar
2 teaspoons grated fresh ginger
½ cup soy sauce
¼ teaspoon freshly ground pepper
2 tablespoons fresh lemon juice
1 large clove garlic, minced
2 tablespoons Worcestershire sauce
2 tablespoons salad or canola oil

Combine all ingredients, except the meat, in a baking dish; mix thoroughly. Add the steaks, cover, and refrigerate at least 6 hours or, for better results, overnight. Lift them from the marinade and drain, saving the marinade for grilling.

Preheat the grill to medium-hot.

Place steaks on the grill so they are 4 to 6 inches above the fire. Turn the steaks over once, and baste with the reserved marinade.

Total grilling time will be about 6 to 8 minutes. For 1-inch-thick steaks grill 6 minutes for rare and about 8 minutes for medium. *Do not cook past medium or 130°F.*

Rib Eye Steaks Topped with Burst Cherry Tomato Wild-Mushroom Sauce

Grassfed beef usually benefits from a few cooking tips. In general, grassfed beef will cook more quickly than other types of beef. When grilling, first sear the meat over high heat, and then move it to a cooler part of the grill to finish cooking. Be careful not to pierce the meat when turning or moving it because the beef will lose some moisture.

Nothing shows off the natural and clean flavor of grassfed beef like a thick, juicy steak, plain and simple, but here we dress up a rib eye steak, one of our favorite cuts, with a lively sauce.

serves 4

for the sauce:

¼ cup extra-virgin olive oil
1 pint red cherry tomatoes, washed
1 pint yellow cherry tomatoes, washed
2 cups sliced mixed wild mushrooms, such as shiitake, oyster, cèpes, or morels
2 medium garlic cloves, minced
Pinch red pepper flakes
1 teaspoon minced fresh oregano
Kosher salt to taste

first make the sauce:

Heat the olive oil in a large skillet over medium-high until very hot. Add all the cherry tomatoes and roll occasionally to heat evenly. Bring to a simmer, cover with a lid, and cook about 8 to 10 minutes, shaking the pan or stirring once.

Remove the lid, add the mushrooms and if tomatoes have not burst, continue cooking, pressing gently with the back of a spoon to mash. Add garlic, red pepper flakes, oregano, and salt to taste.

Lower heat; continue to simmer another 5 minutes or until mushrooms are tender and all ingredients are heated through. Set aside; keep warm.

for the steaks:

- 4 rib eye steaks, about 8 ounces each
- Kosher salt and cracked peppercorns to taste

Heat the grill to medium-high.

Sprinkle one side of the steaks with the salt and cracked peppercorns. Lay steaks on grill at a 45-degree angle. Cook 3 to 4 minutes and then turn 90 degrees to create crosshatching. After another 3 to 4 minutes, flip steaks over and repeat salt and pepper process. Grill until rare or medium-rare when a meat thermometer registers 125°F, about 12 to 15 minutes.

Place the steaks on serving plates and blanket with the burst cherry tomatoes and mushroom sauce.

Save the extra sauce for the next day. Don't worry. It goes quickly.

Grilled Rib Eye Steaks with Roasted Vegetables and Balsamic-Glazed Onions

Complete credit here goes to chefs Stu Stein and Mary Hinds, coauthors of The Sustainable Kitchen *(New Society Publishers, 2004). They follow the 100 percent pastured system and use only the freshest local ingredients. The following recipe, adapted from their cookbook, comprises a full meal executed in brilliant orchestration on the grill.*

serves 2

for the steaks:

1 cup extra-virgin olive oil
2 medium cloves garlic, finely chopped
Kosher salt and freshly cracked peppercorns, to taste, divided use
4 10-ounce beef rib eye steaks, trimmed of excess fat

Preheat grill to medium-high.

Combine the oil and garlic. Set half the oil mixture aside for the steaks and half for the vegetables. Brush remaining oil mixture on steaks. Sprinkle steaks completely with the salt and peppercorns.

Lay steaks on grill at a 45-degree angle. Grill 3 to 4 minutes, and then turn 90 degrees to create crosshatching. After 3 to 4 minutes, flip steaks over, and repeat process. Grill until rare or medium-rare when a meat thermometer registers 125°F, about 12 to 15 minutes.

for the vegetables:

1 large eggplant, cut in half lengthwise and then into ¼-inch-thick slices
1 medium green zucchini, cut into ¼-inch-thick slices
1 medium yellow squash, cut into ¼-inch-thick slices
1 red pepper, left whole
1 yellow pepper, left whole
Kosher salt

Place eggplant in a large colander set over a large bowl; liberally sprinkle with salt. Allow eggplant to sit and give off moisture, about an hour. Wash the eggplant free of salt and pat dry with a towel.

In a large bowl, toss all the vegetables with half the oil mixture; season with remaining salt and pepper. Arrange eggplant, zucchini, squash, and peppers on the grill at a 45-degree angle. Cook 3 to 4 minutes; turn vegetables 90 degrees to create crosshatching. After 3 to 4 minutes, flip over and repeat process. Make sure the peppers brown evenly on both sides.

Remove the vegetables from the grill, set aside, and keep warm. Place peppers in a paper bag, close the top, and steam 10 minutes to loosen their skins, and then scrape or peel off the charred skin. Slice peppers in half, remove seeds, and cut into ½-inch-wide strips.

for the onions:

2 medium yellow onions, thinly sliced lengthwise
1 tablespoon unsalted butter
4 tablespoons balsamic vinegar
Hot pepper jelly
Kosher salt and cracked peppercorns to taste

In a medium-sized sauté pan, melt butter over low-to medium-heat. Add sliced onions and sauté, stirring occasionally, until onions are soft and golden brown, approximately 20 minutes. Add vinegar. Bring to a boil, scrape the bits from the bottom of the pan, and cook until the liquid reduces by one third—about 5 more minutes. Add hot pepper jelly and season with salt and peppercorns. Remove from heat and place in a serving bowl to pour over steaks and vegetables.

Petite Tender Steaks with Portobellos and Yukon Golds

Adapted from Dana Keig and Lindauer Farms, Courtesy of Sustainable Table®

We first saw this recipe using rib eye steaks, which are pretty hard to mess up. We decided to test the shoulder tender, a lesser-known cut, and to sear it on a cast-iron skillet or griddle over an open flame.

This small, individual muscle in the shoulder area is a superior cut that rates high in flavor (similar to beef tenderloin), texture, and easy preparation. Its profile makes it a premium cut that can be used in numerous applications.

Here is a plus: After making the simple marinade, you have coordinated grilling directions for the entire meal of meat, potatoes, and a veggie. Timing is one of the biggest challenges in outdoor cooking, especially as steaks of this size cool so quickly. Just add a garden salad to round out the feast.

serves 2

2 petite tender steaks, about 6 ounces each
Kosher salt and freshly ground pepper to taste

for the marinade:

3 large shallots
6 tablespoons balsamic vinegar
3 tablespoons sugar
4 tablespoons soy sauce

Season the steaks with salt and pepper and place in a large plastic bag. Chop shallots, place in a bowl, and mix well with the vinegar, sugar, and soy sauce. Pour 2/3 of the marinade over the bagged steaks. Marinate steaks in the refrigerator at least 2 hours and up to 2 days.

for the potatoes and mushrooms:

- 6 medium Yukon Gold potatoes
- 1 pound Portobello mushrooms, stemmed
- Extra-virgin olive oil

Prepare the grill by heating to medium-high over half the grill.

Wash potatoes, prick with a fork, and rub with olive oil. First, place potatoes on the heated half of the grill. Total cooking time for potatoes will be close to 35 minutes. Meanwhile, clean mushrooms with a soft brush and cut into ½-inch slices. Pour the reserved marinade over mushrooms.

for grilling the steaks and mushrooms:

Five minutes after placing potatoes on grill, add steaks for 4 to 5 minutes per side for medium rare.

When done, transfer steaks to a cutting board and let rest, tented under foil, for about 8 minutes (potatoes are still on grill). Meanwhile, add mushrooms to grill and brown on each side. When they are done, potatoes will be done, too.

Seasoned Sirloin with Tuscano Sauce

Originally, this recipe was time-consuming and absolutely cheffy, meaning it used techniques few people were likely to try at home. It was like many of the recipes I worked with while handling the food styling for the television show World Class Cuisine *in Italy. This modified version will appeal to those of us who want beautiful, nutritionally dense dishes prepared simply. If you have spent any amount of time in Italy, you will find that the Italians' natural joy of life seeps into all aspects of their cooking. Things like precise measurements are not so very important. So relax, savor the ingredients, and go with the amounts that please you. Radicchio is particular to Tuscany and can be a specialty item; we also roasted escarole along with it. If you can find a rich, red Umbrian wine, go with that, but if not, a nice Chianti will suffice.*

serves 4

for the tomatoes and radicchio:

12 ripe cherry tomatoes
2 Trevigiano radicchio (the long, not round, radicchio) or a small head of escarole
Extra-virgin olive oil, about ¼ cup, divided use
Kosher salt to taste

Preheat oven 300°F.

Cut the tomatoes in half horizontally and the radicchio into 4 lengthwise bunches. Sprinkle with oil and salt, place on a rimmed baking sheet, and roast in the oven for 1 hour.

for the Tuscano sauce:

Sea salt and freshly ground pepper
Fresh sage and rosemary sprigs
4 medium shallots, finely chopped
2½ cups Sagrantino (an Umbrian red wine)
3 tablespoons balsamic vinegar
1 large garlic clove, chopped

Heat the remaining olive oil in a small saucepan over medium-high heat. Add the herb sprigs and shallots and sauté quickly to brown. Mix in the wine and balsamic vinegar and reduce until the mixture becomes syrupy, about 15 minutes. Cool, remove herb sprigs, add garlic, and reserve.

for grilling the steak:

1 2-pound sirloin steak, about 1½-inches thick
4 slices pancetta or nitrate-free bacon, thinly sliced

for aromatic seasoning:

1 teaspoon sea salt flakes
1 sprig each fresh wild fennel, sage, rosemary
Lemon peel from ½ medium lemon
Freshly ground pepper to taste

Heat the grill to high on one side.

Place a cast-iron skillet over the heat and let it get very hot. Brown all sides of the steak in the skillet; leave in pan and cover the top of the meat with the bacon. Close the grill lid and continue cooking until the meat registers 125°F on a meat thermometer placed in the center. Remove the meat from the pan and grill and, tented with foil, let rest on a cutting board about 8 minutes.

to serve:

Prepare the aromatic seasoning (sea salt flakes, wild fennel, sage, rosemary, lemon peel, all chopped together).

Freshly ground pepper to taste

Thinly slice the steak, and drizzle with the reduced Tuscano sauce. Add the aromatic seasoning and freshly ground pepper. Arrange the tomatoes, radicchio, and escarole around the meat slices, and serve. Offer more sauce on the side.

Bourbon Rib Eye Steaks Flamande

James Beard was the pioneering father of outdoor grilling and his knowledge of meat was legendary. He gave us this simple but dramatic flambé technique. Make sure you have your camera ready because you must capture the fun.

P.S. We call for a bit more bourbon than may seem necessary because some will be lost one way or another.

serves 2

2 10-ounce beef rib eye steaks, trimmed of excess fat
Kosher salt and freshly ground pepper to taste
8 sprigs fresh rosemary
½ cup bourbon

Preheat the grill to medium-high.

Rub steaks on each side with salt and pepper. Grill steaks 3 to 4 minutes a side until just rare or medium-rare.

Remove the steaks to a plate lined with a bed of fresh rosemary sprigs; cover with the smaller rosemary sprigs.

In a very small pan, heat the bourbon to only mildly warm. Light the bourbon in the pan and pour the flames over the rosemary and steaks. Voila! A masterpiece!

Grilled Bison Satay

I have prepared a lot of pastured bison, but when a new Annapolis-based restaurant named Level opened serving small plates of local specialties, the bison sampling was outstanding. The satays were perfectly prepared with nicely caramelized crusts and juicy interiors.

serves 4

- ½ cup tamari (low-sodium soy sauce)
- 1/3 cup fresh orange juice
- 1/3 cup honey
- 1 tablespoon toasted sesame oil
- 1 tablespoon fresh lime juice
- 2 tablespoons grated fresh ginger
- 3 medium cloves of garlic, minced
- 1 pound bison steak, partially frozen, then sliced into ¼-inch-thick slices
- 12 bamboo skewers, soaked in cool water
- 2 medium scallions, minced

In a nonreactive bowl, combine the tamari, orange juice, honey, sesame oil, lime juice, ginger, and garlic. Add the bison slices and marinate 2 to 4 hours. Remove slices from the marinade and thread onto the skewers.

Heat grill to high. Sear the satays about 3 minutes per side so that the meat is caramelized, but still medium-rare inside.

Serve hot.

Grilled Tri-Tip Roast with Mojo Sauce

This fabulous recipe from Sunset *magazine uses one of the often-leftover cuts of meat you might be able to get from a farmer. This recipe is perfect with similar cuts of beef or even pork. Make absolutely sure not to overcook large cuts of meat, even if you think they are too rare. Meat continues to cook once it comes off the grill and it has a lot more character when rare.*

serves 6

for the roast:

1 beef tri-tip roast (the bottom portion of the sirloin), 1½ to 2 pounds
¼ cup plus 1 tablespoon extra-virgin olive oil, divided use
2 teaspoons freshly ground toasted cumin seeds, divided use
1 teaspoon dried oregano
1¾ teaspoons salt, divided use
1 teaspoon freshly ground pepper, divided use

Rinse and dry the roast. Rub with 1 tablespoon of the olive oil. Mix together 1 teaspoon cumin, dried oregano, 1¼ teaspoons salt, and ½ teaspoon pepper; massage into the meat and let season overnight in the refrigerator.

for the sauce:

3 tablespoons fresh lime juice
½ cup fresh orange juice
1 teaspoon minced fresh oregano
2 tablespoons garlic, minced
1 small onion, chopped and caramelized

In a blender, whirl juices with remaining cumin, fresh oregano, garlic, ½ teaspoon salt, ½ teaspoon pepper, and the remaining olive oil. Stir in the caramelized onion, and season to taste with additional salt and pepper.

Playing with fire

Here is a handy tip from Margo True, the food editor at *Sunset* magazine. In the July 2008 issue, she writes: "Play with your fire. After years of living in the East, where tri-tip is as common as zebra, I had lost my tri-tip grilling savvy, especially over charcoal. On my first attempt, the coals, perfect at first, dwindled fast, and the meat took ages to cook. Next I tried grooved coals that burned as hot as a blowtorch. Then I remembered: play with the fire to get the right heat. I created a coal-free zone to give me medium heat, and moved the tri-tip there. Later, I pushed it back, chasing medium. At the end, I let the breeze fan the coals into a final burst. Cooking over live fire is like driving a stick shift. It feels good to be back in gear."

for grilling the roast:

Heat the grill to medium-high.

Grill meat, turning to brown evenly and, on charcoal, moving it to wherever the heat is medium-high, until a thermometer inserted in the thickest part reads 125° to 130°F for medium-rare, 23 to 25 minutes. Let meat rest 10 minutes, tented with foil, then cut across the grain into very thin slices. Serve with the sauce.

Asian-Rubbed Flank Steak and Sesame Soy Sauce

The Five-Spice Rub in this treatment is intriguing because the elusive star anise has a flavor most people love but cannot pinpoint. This warm and fragrant spice blend, common in Chinese cuisine, can be found in high-end or specialty supermarkets. Try making your own with our recipe in the rubs chapter (page 204). The toasted sesame oil is easy to find these days, however I have found that Asian groceries carry the most authentic, robust product. The sauce can be kept in the refrigerator up to one month.

serves 4 to 6

1 beef or bison flank steak, about 24 ounces
2 tablespoons Five-Spice Rub
1 tablespoon vegetable oil
Kosher salt

Just before lighting the fire, rub the steak with the Five-Spice Rub, about 1 tablespoon per side.

Preheat the grill to medium-high.

for the sesame soy sauce:

makes about 1 cup

1/3 cup toasted sesame oil
1 tablespoon minced fresh ginger
1 large clove minced garlic
1/3 cup fresh orange juice
¼ teaspoon Five-Spice rub
¼ teaspoon red pepper flakes, optional
2 tablespoons balsamic vinegar
2 tablespoons soy sauce

for the steak:

When the fire is hot, drizzle both sides of the flank steak sparingly with the vegetable oil. Grill the steak 8 minutes, flipping it a couple of times during cooking. Be sure to keep the thicker part over the hotter area of the fire and the thinner, tapered end over the cooler area. It should be rare to medium-rare at this point or about 125°F on a meat thermometer. Transfer the steak to a carving surface and let rest 5 minutes.

to serve:

To carve, cut very thin slices across the grain and on a 45-degree angle. Drizzle with the sesame soy sauce and serve more on the side.

Beef or Bison Kebabs and Potato Brochettes

Try this recipe with either beef or bison. The rib eye cut of bison is surprisingly lean and it yields tender results when grilled. Bison has a deep, slightly richer flavor than beef and it stands up well to the rosemary. Serve these kebabs with roasted mixed bell peppers and a lush garden salad to round out your meal.

serves 2 to 4

for marinating the steak and potatoes:

¼ cup extra-virgin olive oil
1 tablespoon fresh rosemary, finely chopped
1 tablespoon freshly ground pepper
2 large garlic cloves, minced
2 tablespoons fresh shallots, minced
1 pound rib eye steak, trimmed, and cut into 1¼-inch cubes
½ teaspoon kosher salt
4 3-inch red bliss potatoes, scrubbed and cut in half
Olive oil cooking spray

Combine the first five ingredients in a large bowl. Add the beef and the potatoes; toss well to coat. Cover and refrigerate 45 minutes.

for grilling the brochettes:

Heat the grill to medium-high.

Remove meat and potatoes from the marinade; heat remaining marinade to just a boil to kill any bacteria. Reserve for basting sauce.

Thread meat and potatoes evenly onto each of 4 12-inch skewers.

Place the skewers on a grill rack coated with cooking spray; grill 3 minutes on each side or until medium-rare. Serve immediately.

Yakimono Woven-Beef Tenderloin Skewers

Woven meat skewers can be found in nearly every cuisine. In this recipe we used the Japanese version as I had worked with yakimono when I helped with the concept and development of Entertaining on the Run, *by Marlene Sorosky. Skirt steak strips or even sirloin can be used instead of the tenderloin. These yummy and garlicky slices get their kick from tenderizing, spicy Dijon mustard. The balsamic vinegar not only tenderizes, it adds a touch of sweetness.*

serves 8

4 medium cloves garlic, finely chopped
¼ cup Dijon or spicy mustard
2 teaspoons smoked Spanish paprika
¼ teaspoon kosher salt
¼ teaspoon freshly ground pepper
1 tablespoon low-sodium soy sauce
2 tablespoons balsamic vinegar
1 2-pound beef tenderloin, halved and cut into 1-inch slices
8-inch wooden skewers, soaked in cold water for 30 minutes

for the glaze:

Whisk together the first 7 ingredients in a small bowl. Cover and let sit at room temperature 30 minutes before using.

to grill the beef:

Heat the grill to high.

Skewer 2 pieces of the beef onto 2 skewers so that the meat lies flat. Brush the meat liberally on both sides with the glaze. Grill the meat 2 to 3 minutes per side until golden brown and cooked to medium-rare, brushing with the remaining glaze while grilling. Remove from grill and serve.

Java-Pasted Brisket

There are many different ways to use coffee to enhance grilled foods. The acid in coffee acts as a tenderizing agent while delivering a deep and robust flavor to any slow-roasted meat. Java paste is a wet rub, a paste similar to the consistency of the spice blends of India. Use the rub on other meats like poultry, pork, and just about anything you want to try. How about that whole turkey?

serves 8 to 10

4 tablespoons finely ground dark-roast coffee beans
¼ cup canola oil
2 tablespoons brown sugar
4 large cloves garlic, finely chopped
2 tablespoons Worcestershire sauce
1 teaspoon freshly ground pepper
1 teaspoon onion powder
1 teaspoon freshly ground cumin seed
1 teaspoon freshly ground fennel seed
½ teaspoon kosher or other large-flake salt
1 teaspoon bottled hot sauce
1 center-cut beef brisket, 5 to 6 pounds, trimmed so there's still a ¼-inch-thick layer of fat on top

In a small bowl, stir together these 11 seasoning ingredients. Reserve approximately ¼ of the rub for the basting sauce. I like to place the brisket in heavy foil and then put that in a roasting pan (even if it's the disposable kind). With your hands, generously pat the paste on the entire surface of the brisket. Let the paste-covered brisket season overnight in the refrigerator, covered.

for grilling the brisket:

Preheat the smoker or half the grill to medium-high.

Place the brisket, still in the foil and pan, over the heat source and close the lid for about 20 minutes to allow the meat to brown. Reduce the heat to medium-low, move the brisket to the indirect side of the grill, wrap the foil tightly over the brisket to seal, and continue cooking for 4 to 5 hours on low. Make sure to check at least once every hour to see that it's not cooking too quickly.

for the basting sauce:

¾ cup apple cider
½ cup water

About 1 hour into the grilling time, open the foil and pour all pan juices from the meat into a medium-sized saucepan. Add the reserved paste mixture along with the apple cider and water. Bring this to a rolling boil and cook, stirring, for 5 minutes; begin basting the brisket every hour or so by opening the foil and brushing with the sauce, always covering again with the foil after basting.

After 4 hours, check to make sure the meat is very tender, moist, and almost falling apart. If not, continue grilling for what may be another hour or two.

Remove the meat from the grill and let it rest for 10 minutes, loosely covered with foil. Place on a cutting surface and thinly slice across the grain with a very sharp carving knife. Save any juices. Transfer the sliced meat to a platter, spoon juices and any remaining basting sauce over the slices.

Tapas, Ground-Beef Albóndigas with Spanish Sauce

Tapas are always alluring but especially so if you have visited Spain, where they originated. And who doesn't love a healthy meatball smothered in a zesty red sauce chock-full of tomatoes?

The meatballs—albóndigas—are small and difficult to sear on the grill rack, so use a greased sheet of heavy foil or an additional grilling pan for these smaller pieces of food.

serves 4 as an entrée, 8 as hors d'oeuvres

for the meatballs:

2 teaspoons extra-virgin olive oil
1 small onion, minced
1 large garlic clove, minced
1 pound pastured ground beef
1 egg, beaten
1/3 cup plain fresh bread crumbs
2 tablespoons fresh lemon juice
½ teaspoon freshly ground cumin seed
1 teaspoon Dijon or spicy mustard
2 tablespoon fresh parsley, minced
Kosher salt and freshly ground pepper to taste

Warm the olive oil in a sauté pan over medium-high heat. Sauté onion and garlic until soft and let cool. Place the ground beef in a bowl and add the egg, bread crumbs, lemon juice, cumin, mustard, parsley, salt and pepper, and then the sautéed onion and garlic. Mix well and roll into very small meatballs—about 1 inch in diameter. Cover the meatballs and place in the refrigerator to let cool 1 hour.

for the spicy Spanish sauce:

- 2 tablespoons extra-virgin olive oil
- 1 medium onion, finely chopped
- 1 pound fresh tomatoes, finely diced with juices, or 1 14-ounce can of puréed tomatoes
- 2 tablespoons dry sherry
- 1 teaspoon Spanish paprika
- Pinch of sugar
- ½ teaspoon cayenne or to taste
- Kosher salt and freshly ground pepper to taste
- Freshly chopped parsley, for garnish

Place the olive oil in a sauté pan over medium-high heat. When hot, add the onion and sauté for 5 minutes. Add the tomatoes, sherry, paprika, sugar, cayenne, salt, and pepper, and simmer 20 minutes. Now add the garlic and simmer 5 more minutes to infuse the flavor.

for grilling the meatballs:

Preheat heat the grill to high.

Grill the meatballs 5 to 10 minutes to sear, turning frequently until evenly browned. Place them in the hot sauce, heat another 10 minutes, and serve hot topped with a sprinkle of parsley.

Smoked-Beef, Farm-and-Pasture Minestrone

This recipe got its name from the freshly picked produce and smoked grassfed-beef marrow bones we used to make the rich and soulful beef stock. Try this one when you are in the mood for a dinner that is cozy, delicious, and healthy. The chop-and-drop method makes it easy cooking for a weeknight. Even if there is only one or two of you, make the whole thing because the leftovers only get better.

serves 12

2 large beef marrow bones, smoked

to smoke the beef marrow bones:

Preheat the grill to medium-high.

Put the bones in a rimmed baking pan lined with heavy aluminum foil and place over the heat source; close the grill lid for about 20 minutes for browning. Now reduce the heat to medium-low, and move the pan to the indirect side of the grill. Place some soaked fruitwood or herbs "on foil" over the direct flame until smoking. Continue adding smoke fuel to the flames as needed and smoke the marrow bones about 1½ hours until nicely browned, with any meat pulling away from the bone.

for the soup stock:

8 cups (2 quarts) water
2 bay leaves

Heat a medium-sized soup pot over medium-high and add the bay leaves and beef marrow bones. When the water comes to a boil, reduce heat and simmer 2 hours to make a very rich stock.

At this point, cool the stock, remove the bones and meat with a slotted spoon. Chop the meat and save for them for the soup (along with the stock). Discard the bones.

If making for another use, cool and freeze the stock for up to 4 months.

for the minestrone:

- 2 tablespoons extra-virgin olive oil
- 1 small hot chile pepper
- 4 medium garlic cloves, minced
- 2 medium onions, chopped
- 2 medium carrots, peeled and diced
- 1 pound Yukon Gold potatoes, scrubbed and quartered
- 2 celery ribs, chopped with greens
- Kosher salt and freshly ground pepper to taste
- 2 stems of fresh rosemary
- 6 fresh sage leaves, thinly sliced
- 1 medium zucchini, diced
- 4 to 5 cups (a small bunch) kale or chard, tough ends and veins trimmed, coarsely chopped
- 1 15-ounce canned canellini beans (also called white or red kidney beans), drained
- 1 cup fresh tomatoes, diced
- 1 14-ounce can fire-roasted diced tomatoes
- 1 small piece, 2 × 3 inches, Parmigiano cheese rind

Heat a medium-sized soup pot on medium-high and add the olive oil. Add the chile pepper, garlic, onions, carrots, potatoes, and celery. Cook 5 to 6 minutes. Season with salt and pepper, add the rosemary and sage, and mix well. Add the zucchini and greens; stir until all greens wilt down, 2 to 3 minutes.

Add the beans, tomatoes, reserved stock, reserved meat, and cheese rind; cover the pot with a lid, and bring soup to a slow boil. Cook, covered with the lid, until all vegetables are tender but not mushy—just about 20 minutes. Remove the pot from the heat. Remove the rind and the now bare rosemary stems (the leaves fall into the soup as it cooks). Adjust the salt and pepper to taste.

to serve:

- Additional Parmigiano, grated
- Crusty bread, for mopping

Ladle the soup into shallow bowls and top with grated cheese.

Note: The soup can be cooled and frozen for up to 3 months.

Smoky Bison Meatloaf

Inspired by Good Eats, *Alton Brown's television show, we love the addition of something as healthy as rolled oats to absorb liquid and retain tenderness. A food processor makes quick work of chopping. Do not add the meat to the food processor; stir in manually. Ground meat should be handled as little and as lightly as possible. In fact, don't even squeeze the final mixture.*

serves 6 to 8

for the meatloaf:

1 cup old-fashioned rolled oats
½ teaspoon freshly ground pepper
Dash cayenne
½ teaspoon barbecue spice blend of choice. For suggestions see Chapter 10, Rubs, Mops, Marinades, Brines, and Sauces (page 193)
1 teaspoon dried oregano
½ medium onion, coarsely chopped
1 carrot, peeled and coarsely chopped
3 medium cloves garlic
½ red bell pepper, seeded and coarsely chopped
1½ pounds ground bison
1½ teaspoons kosher salt
2 eggs plus 1 egg white

for the glaze:

¾ cup chile sauce (available in the ketchup aisle of the supermarket)
¼ cup pepper jelly, melted
1 teaspoon ground cumin
Dash Worcestershire sauce
Dash bottled hot pepper sauce
1 tablespoon honey

In a food-processor bowl, combine oats, pepper, cayenne, barbecue spice, and oregano; pulse to a fine powder, and then move to a large bowl.

Combine the onion, carrot, garlic, and red pepper in the food processor bowl. Pulse until the mixture is finely chopped, but not puréed.

Add the ground bison and the mixed vegetables to a large bowl; season with salt. Beat the eggs, add to the meatloaf mixture and combine thoroughly, but avoid squeezing the meat.

to grill the meatloaf:

Heat the grill to medium.

Pack the meatloaf into 2 small loaf pans, 3 × 6 inches. Place the pans on the indirect heat side of the grill and close the lid. Check the meat temperature after 20 minutes. Medium-well is the best doneness for a moist bison meatloaf so cook close to 25 minutes, total, but rely heavily on the meat thermometer. It should read 140°F.

for the glaze:

Combine ingredients. Brush onto the meatloaf after it has been cooking about 10 minutes and again several minutes before it is finished.

Rolled Stuffed Flank Steak

Either bison or a beef steak works beautifully in this recipe and we suggest trying each one. The ingredients for the filling are interchangeable, so use black olives, mushrooms, or sun-dried tomatoes, or any other tasty morsel you prefer.

serves 6

for the marinade:

- ¼ cup soy sauce
- ¼ cup balsamic vinegar
- 2 medium cloves garlic, minced
- ½ teaspoon cracked black pepper
- 1 flank steak, about 1½ pounds

for the filling:

- ⅔ cup celery, finely chopped
- 1 roasted bell pepper, patted very dry and diced
- 1 cup pitted green olives, coarsely chopped
- 2 tablespoons fresh parsley, minced
- 1 medium clove garlic, minced
- 2 cups fresh baby spinach, stems removed
- Kosher salt and freshly ground pepper to taste

Using a meat tenderizer or a fork, prick the steak evenly over the entire surface; turn over and repeat. In a plastic zipper bag, combine soy sauce, balsamic vinegar, garlic, and black pepper. Add the steak and, turning occasionally, marinate, refrigerated, for several hours or overnight.

When ready to prepare the meal, preheat the grill to medium-high.

In a medium bowl, mix the celery, roasted pepper, green olives, parsley, and garlic. Place the steak on a work surface and, first, spread the fresh spinach over the meat, then the vegetable mixture, and press down to compact. Roll the meat up from its narrowest end and tie at 2-inch intervals with cotton butcher's string.

Brush the meat with the marinade and sprinkle with salt and pepper. Place on the grill over direct heat and turn to sear on all sides, about 8 minutes. Reduce the grill heat to low and continue grilling, covered with the lid, another 18 to 22 minutes or until a meat thermometer reaches about 125°F for rare. Be sure to turn the stuffed flank once during the lower grilling heat. When done, let rest 10 minutes, loosely tented with foil. Carve into ½-inch thick slices, remove the butcher's string, and serve.

Basic Meat Stock

Stock, an elemental ingredient for consommé, soups, and sauces, develops from slowly simmered meat with or without bones, aromatic vegetables, herbs, and spices in a water base. This strained elixir becomes the base for soups and sauces and is often used for thinning. Roasting the meat and bones first produces caramel-colored stocks with rich, flavorful profiles. Although we direct this recipe toward beef, bison, pork, lamb, or goat, the foundation is the same for poultry. And you know the old saying: chicken soup cures a cold.

The bones from meat and poultry produce stocks that are high in calcium and other minerals that our bodies need. Bones from grassfed animals are the key to rich, gelatinous stocks. We can derive the nutritional benefits of pasture greens through the bones of grazing animals. Boney bones, such as marrow and knucklebones, create gelatin and infuse stocks with minerals; meatier bones, such as chuck ribs and neck bones, bring rich color and flavor. The combination of these types of bones is the secret to a stock's ideal flavor and nutrition. Reduced stocks, thickened by rapid boiling, form the foundation for numerous sauces.

makes about 8 cups

for browning bones:

Restaurant chefs use this browning technique for the bones because it is straightforward and does not need the constant attention and turning necessary when browning in a pot. However, because this is a grill book, we also suggest browning the bones on the grill.

for the stock:

8 cups cool water
3 pounds meaty beef and boney bones, such as ribs, shin, neck, and marrow
1 medium onion, peeled and quartered
2 medium carrots, peeled
3 medium ribs celery
2 large cloves garlic, crushed
2 sprigs fresh thyme
1 bay leaf
½ teaspoon kosher salt

Preheat oven to 400°F.

Place the bones on a sturdy baking sheet and roast 45 minutes or until the bones brown and the meat is crisp.

Place a 10-quart soup pot on high heat, add the water, beef bones, accumulated juices, along with any browned bits you have scraped and loosened from the bottom of the roasting pan. Bring the stock to a simmer and, using a ladle or large spoon, skim off any impurities or foam that rise to the surface.

Add the onion, carrots, celery, garlic, thyme, bay leaf, and salt. Cook uncovered, at the barest possible boil, with just a few bubbles breaking the surface, stirring occasionally, for 2½ hours.

Remove from the heat and let the stock rest for 1 hour or until cool. Move stock into a large container for cooling while straining it through a fine-mesh strainer, a china cap strainer, or a colander double-lined with damp cheesecloth.

Transfer the cooled stock to airtight containers, and refrigerate for up to 3 days or freeze up to 3 months.

5

Pork

PASTURED PIGS NEED PASTURE . . . AND MORE

Raising hogs on pastures drenched in sunshine and sweetened with fresh air has so many benefits. If you're not convinced, consider the alternative: for hogs, that means confinement in crowded industrial facilities with unsanitary buildups of waste and exposure to noxious dust and odors. Pasture-raised hogs are free to roam, root, and graze healthy clovers and grasses, which supplement their diets. Doing so reduces the dependence on the two primary ingredients that make up a conventionally raised swine's diet: corn and soybeans. Producing corn- and soybean-based feeds takes considerable amounts of diesel fuel, pesticides, and fertilizers, all of which have an impact on the environment.

Unlike cows, hogs are not ruminants; they are not able to obtain even half the nourishment they need to be healthy from pasture alone. Sows and piglets benefit most from the nutrition they get from fresh, green pastures. In fact, sows can derive nearly 50 percent of the nourishment they pass on to their young from what they find on pasture. Hogs get only about 20 percent of their nourishment from pastures. This means in addition to pasture, a hog's diet must include grains and vegetables. As a result, the nutritional benefits of hog meat, while good, do not include the full range of benefits that come from eating the meat of pasture-raised cows, bison, sheep, and goats. Many farmers who are producing pastured pork let their hogs root and glean in vegetable fields after the vegetable harvests. Hogs that have access to wooded areas will feast on acorns and other nuts, thereby enhancing the flavor and nutritional value of pastured pork. This again speaks to the importance of knowing how the animals you eat are raised.

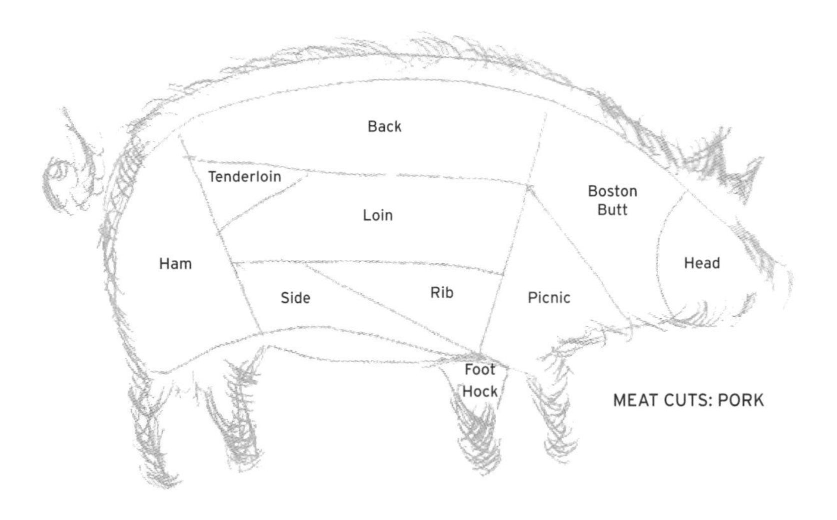

MEAT CUTS: PORK

Pork Loin Chops with Maple Dijon Sauce and Pecans

So simple yet rich and satisfying. I grill year-round as the fresh air and smoky aroma are invigorating enough to rouse the appetite.

serves 4

4 6-ounce pork loin chops, 1¼-inch thick
Kosher salt and freshly ground pepper

for the maple dijon sauce:

2 tablespoons Dijon mustard
2 tablespoon maple syrup
1 medium clove fresh garlic, minced
¼ cup butter, melted
½ cup toasted chopped pecan halves

Combine ingredients

Dry the chops with paper towels and season generously with salt and pepper. Insert a long skewer width-wise through each pork chop.

Heat the grill to medium-high.

Grill chops over direct heat for 6–7 minutes; turn and slather with Maple Dijon Sauce and sprinkle with most of the pecan halves. Close the lid and grill 5-6 minutes more for medium doneness.

Remove the pork chops from the grill, being careful not to spill the sauce and nuts. Place chops on a serving plate and top with a final crowning of the remaining pecans.

Brined Pork Chops with Grilled Nectarine Glaze

Because today's pastured pork is very lean, the meat can easily dry out when overcooked. Using brine—a solution of water, salt, and sometimes sugar and seasonings—adds moisture and flavor. We suggest, especially with pork, to stay with the low-and-slow cooking method. The thickness of the pork chops in this recipe ensures that the meat won't dry out, and the fruit sauce combined with the pork is heavenly.

Most of the nectarine glaze can be made one day ahead and chilled, then warmed over medium heat when adding the grilled diced nectarines.

serves 4

for the brine:

½ cup kosher or sea salt
½ cup packed light brown sugar
2 sprigs plus 1 teaspoon finely chopped fresh lemon verbena, divided use
2 teaspoons whole peppercorns

In a large pot, bring 1½ cups of water to a rolling boil. Remove from heat; stir in the salt, brown sugar, lemon verbena sprigs, peppercorns, and stir until salt and sugar are completely dissolved. Add 1½ cups large ice cubes to chill the brine until very cold. Place the pork chops in the brine and weight down with a plate to keep meat completely submerged. Chill overnight.

for the nectarine glaze:

3 tablespoons olive oil, divided use
2 cups thinly sliced sweet onion such as Vidalia
1 tablespoon apricot jam
2 teaspoons low-sodium soy sauce
2 tablespoons sherry vinegar
1 teaspoon chopped fresh lemon verbena
2 to 3 nectarines, washed, halved, and pits removed

Place 2 tablespoons of the oil in a large frying pan over medium heat. Add onions and sauté, stirring often, until caramelized, about 10 to 15 minutes. Turn the heat to low and add the apricot jam, soy sauce, and sherry vinegar. Stirring often, cook until the mixture caramelizes, about 15 minutes. Stir in the chopped lemon verbena.

While the grill heats, using tongs, add the nectarine halves, cut side down, on direct flame and grill until brown; finely dice and add to the warm glaze.

for grilling the pork chops:

Heat the grill to medium.

4 center-cut pork chops, bone in, 1-inch thick

Remove pork from brine and pat dry. Brush the pork all over with the remaining tablespoon of olive oil.

Grill the pork, covered, turning once, until meat is medium-rare, about 140°F on a meat thermometer. Transfer the pork to a platter, tent loosely with foil, and let rest 5 to 10 minutes. Serve topped with the grilled nectarine glaze.

Tuscan-Style Pork Loin Roast with Smoked Tomato and Olive Sauce

This roast reminds me of many of the experiences I had in Italy while working for World Class Cuisine *on the Discovery Channel, especially the whole baby pig we grilled in a smokehouse, over an open fire, for a wedding banquet. Of course, rosemary and sage are the great herbs of the country. Pastured pork is widely available and widely enjoyed throughout Italy and it marries well with many styles of outdoor cooking. If you have ever purchased a rather large roast, such as a pork loin, you have probably spent a good amount of money: so prepare it as a celebration in itself.*

The pork is best when brined 8 to 18 hours ahead, but overnight is the way to go. Bring the meat to room temperature while you heat the grill.

serves 6 to 8

for the brined pork:

- ¾ cup kosher salt
- ¼ cup packed light brown sugar
- 3 medium cloves garlic, minced
- 3 small sprigs fresh rosemary
- 1 large sprig fresh sage
- 2 cups water
- 1 2½- to 3-pound pastured pork loin roast, trimmed of excess fat

In a 4-quart saucepan, combine the salt, sugar, garlic, herb sprigs, and water. Stir over high heat just until the salt and sugar dissolve. Add 6 more cups of water plus 1½ cups of large ice cubes to cool very quickly and cool to room temperature. Transfer to a large container, add the pork, cover, and refrigerate 8 to 18 hours.

for the herb paste:

- 8 medium cloves garlic, peeled
- ¼ cup fresh rosemary leaves
- ¼ cup fresh sage leaves
- ¼ cup fresh sorrel leaves, if available
- Kosher salt and freshly ground pepper, divided use
- 3 tablespoons extra-virgin olive oil

Put the garlic, rosemary, sage, sorrel, 1 teaspoon salt, and 1 teaspoon pepper in a food processor or large mortar and process until coarsely mixed. Slowly add the oil to form a paste. Save 1 tablespoon of this paste for the sauce; spread the herb paste all over the outer surface of the roast and press firmly to make sure it sticks.

for the smoked tomato and olive sauce:

- 1 tablespoon of the reserved herb paste
- 1 cup smoked tomatoes (page 212)
- 1 cup grape tomatoes, finely diced
- Zest and juice of 1 large lemon
- ¼ cup Kalamata olives, pitted and coarsely chopped
- ½ to 1 cup poultry stock (page 159), or dry white wine for thinning sauce to pourable consistency, if needed

Place all the ingredients in a medium saucepan and heat gently, just until hot. You don't want to overcook this rustic, chunky sauce and lose its fresh flavor. A little warming is all that's needed.

for grilling the roast:

Heat the grill to medium-high for indirect grilling.

Insert the entire pork roast lengthwise on the rotisserie spit and let rotate over the grill, covered, until a meat thermometer inserted near the center of the roast registers 145°F, about 35 to 45 minutes.

If you don't have a rotisserie, set up the grill for indirect grilling. Heat the grill to medium. Place the roast in indirect heat on the grill, and with the lid closed, turning the roast every 15 minutes. The roast should still register about 145°F, about 35 to 45 minutes.

Remove the roast from the grill and transfer it to a cutting board. Let stand for 10 minutes, loosely tented with foil (the roast will continue to cook as it stands), and then slice thinly; serve hot, warm, or at room temperature with the sauce on the side.

Adobo Pork Tenderloin with Pineapple Pico de Gallo

Here we give you the breakdown of the adobo rub that has become so popular. We want to encourage you to make your own blend because store blends are often loaded with preservatives. Keep it on hand and add it to your stock of condiments.

serves 4 to 6

for the adobo rub:

- 6 tablespoons paprika
- 1 tablespoon chile powder
- 3 pinches cayenne
- 2 tablespoons kosher or other coarse salt
- 2 tablespoons freshly ground pepper
- 2 teaspoons ground dried oregano
- 2 tablespoons brown sugar
- 2 pounds pork tenderloin, trimmed

In a small bowl, combine the spices, salt, pepper, oregano, and sugar. Completely rub each tenderloin with the adobo mixture.

for grilling the pork:

Preheat the grill to medium-high.

When hot, add the tenderloin and grill for about 15 to 20 minutes total, rolling periodically to brown evenly on all sides. It should read about 125° to 130°F on a meat thermometer. Let rest covered loosely with foil for 5 minutes; carve in ¼-inch slices and save all the juices to drizzle over the meat.

for the pico de gallo:

¾ cup finely chopped fresh pineapple
2 cups cooked black beans, freshly cooked or canned
3 medium tomatoes, coarsely diced
½ cup red onion, coarsely diced
½ cup green onions, coarsely chopped
½ cup fresh cilantro, finely chopped
1 teaspoon jalapeño pepper, minced
2 tablespoons fresh lemon juice
1 tablespoon chile powder
¼ teaspoon kosher salt
2 cups arugula, loosely packed, for the serving platter

In a medium-large bowl, gently mix all ingredients except the arugula.

to serve:

Place the arugula on a serving platter and top with the tenderloin slices. Serve the pico de gallo over the sliced tenderloin.

Glazed Pork Loin with Corn, Tomato, and Basil Salsa

This combination of flavors showcases summer's bounty with a real outdoor all-American style. The sweet spiciness of the glaze along with the vegetables in the salsa sing together over a platter of grilled meat. Sandwiches of the seasoned pork topped with salsa are divine the next day . . . if there is any left. You may want to double the salsa recipe because it's good enough to just eat by the spoonful.

serves 4

for the glaze:

3 tablespoons fresh lime juice
3 tablespoons honey
2 medium cloves garlic, minced
1 teaspoon kosher salt
2 pork tenderloins, ¾ to 1 pound each

for the salsa:

1 cup fresh plum tomatoes, diced
1 cup freshly cooked corn kernels, cut from the cob
½ cup red onion, minced
3 tablespoons fresh basil, finely chopped
1 tablespoon white wine vinegar
2 tablespoons fresh lime juice
¼ teaspoon kosher salt
1 jalapeño, seeded and minced

Combine the glaze ingredients, then coat the tenderloins thoroughly. Let rest in glaze for 1 hour in the refrigerator.

Meanwhile, make the salsa by mixing ingredients together and hold at room temperature to let season until ready to serve with the pork.

Preheat the grill to medium-high.

Place the tenderloins on the grill and cook, turning frequently, for about 18 minutes, or until an instant-read thermometer inserted into the thickest part reads 150°F.

Transfer to a platter and allow the meat to rest 10 minutes, loosely tented with foil, before carving.

Serve with salsa on the side.

Lacquered-Plus Pork Loin Chops

serves 4

for the marinade and sauce:

- 3 tablespoons toasted sesame oil
- Zest and juice of 1 fresh orange
- ½ cup apricot preserves
- ¼ cup low-sodium soy sauce
- 3 medium cloves garlic, minced

Mix all ingredients in a small bowl.

for the pork and vegetables:

- 4 loin pork chops, 1 inch thick
- 1 pound baby Portobello mushrooms, cleaned and cut in half
- 1 pound sugar snap peas

Place the pork chops in a resealable bag and the vegetables in another. Divide the marinade between the two bags and let season, refrigerated, for 2 hours.

Preheat the grill to medium-high.

Place a flat, cast-iron griddle on the flames, and heat.

Spray the hot griddle with nonstick cooking spray; add the pork chops and sear for 3 minutes on each side to brown. Now, move the griddle to indirect flame and continue cooking the pork for about 8 more minutes until a meat thermometer reaches 150°F for medium.

Sauté the vegetables on the same grill until just crisp-tender.

Remove the pork and vegetables from the grill and serve immediately.

Hoisin-and-Rum-Glazed Pork Tenderloin

Versions of this recipe abound and that's a good thing because it gets very high marks. Note that you are soaking wood chips.

serves 4

1 cup hickory wood chips

for the marinade:

¼ cup Hoisin sauce
1 tablespoon seasoned rice vinegar
2 tablespoons dark rum
2 tablespoons maple syrup
1 tablespoon ginger, peeled, freshly grated
2 teaspoons fresh lime juice
1 teaspoon chile paste with garlic
1 large garlic clove, minced
2 pork tenderloins, ¾ to 1 pound each
Cooking spray

Combine Hoisin sauce and the next 7 ingredients in a small bowl; stir with a whisk. Place the pork tenderloins in the marinade and refrigerate overnight.

to grill the pork:

Soak wood chips in water 30 minutes; drain well.

Heat grill to medium-high.

Place the wood chips on a piece of heavy foil punctured with about three 4½-inch holes to allow the smoke to infuse the meat and put on the grill on top of a flame. Gently shake the tenderloins to remove excess marinade and place them on the grill rack that you have coated with cooking spray; cook 5 minutes. Turn and baste pork with Hoisin mixture; cook 5 minutes; let rest 5 minutes before serving.

Country-Style Pork Ribs with Tequila-Lime Sauce

You can guess this sauce packs a pleasant wallop but it is versatile enough to use on many different meats and poultry to achieve that "land of the sun" quality. Save any extra chunky sauce in jars to give as a gift or just to have on hand. (Keep them refrigerated for one month because tequila acts as a preservative.) For a velvety sauce, place the sauce in a blender and process until smooth.

serves 4

for the tequila-lime sauce:

1 small jalapeño, finely chopped
6 medium cloves garlic, minced
1 medium sweet onion, finely chopped
1 Cubanelle or other mild pepper, finely diced
2 medium tomatoes, finely chopped
2 tablespoons honey
2 tablespoons fresh lime juice
½ teaspoon kosher salt
1 cup golden tequila

Combine all ingredients in a saucepan and simmer about 10 minutes until the vegetables soften. Let cool, and then marinate the ribs in enough sauce to cover. Refrigerate and let soak up flavors for 2 hours to overnight.

for the ribs:

4 country-style pork ribs

Preheat the grill to medium-high.

Make sure the ribs retain some sauce and place over slow coals. Turn every 10 minutes, brushing frequently with sauce. Grill at least 5 to 7 minutes per side until done—use a meat thermometer to read up to 140°F because these ribs are usually pretty thick. Grill about 35 minutes or until done and nicely glazed with the sauce.

Pork Chops Milanese

We've lightened up this Milanese recipe a bit from the fried version and adapted it for the grill. The panko bread crumbs do not overwhelm the flavor of the fantastic grassfed pork. The potatoes are the traditional complement.

serves 6

6 boneless pork chops, cut 1-inch thick
2 free-range eggs, beaten
2 cups Japanese panko bread crumbs
3 tablespoons Parmesan cheese, finely grated
1 lemon, zest only
Kosher salt and freshly ground pepper to taste
Natural spray olive oil
12 cooked new potatoes, for serving

for the sauce:

2 tablespoons fresh Italian parsley, finely chopped
2 tablespoons fresh basil, finely chopped
1 tablespoon fresh sage, finely chopped
1 teaspoon capers, drained and rinsed
Zest and juice from 1 lemon
1 medium garlic clove, crushed to a paste with the edge of a knife
Kosher salt and freshly ground pepper

Preheat the grill to medium-high.

Place the beaten eggs in a bowl. In a separate bowl, mix the panko, Parmesan, and lemon zest until well combined. Season with salt and pepper.

Dip each pork chop in the beaten egg; dredge in the panko mixture until completely coated. Place the coated pork chops on a sheet of heavy foil coated with olive oil spray.

Grill 3 minutes for the first side then turn and grill another 3 minutes. Turn again to the first side, move further from the direct flame, and grill another 8 minutes.

for the sauce:

In a bowl, whisk together the finely chopped herbs, capers, lemon zest, lemon juice, and garlic until well combined. Season with salt and pepper.

To serve, place 1 pork chop on each of 6 serving plates with 2 potatoes alongside each. Drizzle with the sauce.

Jim Tabb's Barbecue Pork, His Rub, and His Honey Mustard Barbecue Sauce

Mr. Tabb was the first authentic pitmaster I ever met, and a truly excellent one at that. He's the real thing. When Jim and I were part of the Smithsonian's Folklife Festival, he hauled his long, black, tubular smoker up from North Carolina to smoke lots of meat for the special events. We all inhaled his masterpiece and, afterward, when we could breathe again, we broke up with laughter from his stories about his run-ins with the Smithsonian Mall security detail and how the FBI followed him and his bomb-like grill for miles.

serves 8

for the barbecue rub:

makes 3 cups

1¼ cups firmly packed dark brown sugar
1/3 cup kosher salt
¼ cup granulated garlic
¼ cup paprika
1 tablespoon chile powder
1 tablespoon freshly ground red pepper
1 tablespoon freshly ground cumin
1 tablespoon lemon pepper
1 tablespoon onion powder
2 teaspoons dry mustard
2 teaspoons freshly ground pepper
1 teaspoon freshly ground cinnamon

Combine all ingredients. Store in an airtight container.

for the pork:

1 6-pound pork shoulder roast, bone in
1 cup barbecue rub
Hickory wood chunks
Apple juice, about 1 cup for spraying pork
Small spritzer for liquid

Trim fat on the pork shoulder roast to about 1/8-inch-thick. Sprinkle the pork evenly with barbecue rub, rub thoroughly into the meat, wrap tightly with plastic wrap, and chill 8 hours.

Discard the plastic wrap. Let pork stand at room temperature 1 hour.

Start soaking hickory chunks in water for 1 hour.

Prepare the smoker according to manufacturer's instructions, bringing internal temperature to between 225° and 250°F and maintain temperature for 15 to 20 minutes.

Drain wood chunks, and place on coals.

Place pork on lower cooking grate, fat-side up.

Spritz pork with apple juice each time you add charcoal or wood chunks to the smoker.

Smoke the roast, maintaining the temperature inside the smoker between 225° and 250°F for 6 hours or until a meat thermometer inserted horizontally into the thickest portion registers 170°F. Remove pork from the smoker, and place on a sheet of heavy-duty aluminum foil; spritz with apple juice, return to smoker, and smoke 2 hours longer or until thermometer inserted horizontally into the roast registers 190°F. Remove pork from smoker, and let stand 15 minutes. Remove bone and pull pork into bite-size shreds.

for the honey-mustard barbecue sauce:

makes 1/2 cup

This is a thick, hearty sauce that really sticks. If you prefer a thinner sauce, add ¼ cup water. Whichever way . . . slather it over that irresistible pulled pork.

- 1 bacon slice, diced
- 1 small onion, diced
- 1 garlic clove, minced
- 1 cup apple cider vinegar
- ¾ cup Dijon or spicy mustard
- ¼ cup firmly packed brown sugar
- ¼ cup honey
- 1 teaspoon freshly ground pepper
- 1 teaspoon Worcestershire sauce
- ¼ teaspoon freshly ground red pepper

In a medium-sized saucepan over medium-high heat, cook the bacon until crisp; remove and drain on paper towels, reserving drippings in the saucepan.

Sauté the onion and garlic in the hot drippings about 3 minutes or until tender. Stir in bacon, vinegar, and remaining ingredients; bring to a boil. Reduce heat, and simmer, stirring occasionally, 10 minutes. Store in refrigerator for up to 1 week.

Fusion Spicy Shortcut Ribs

Folks flock to these very spicy ribs, but they can't always identify the heat as Sriracha, a bright orange-red Thai paste that gets its color from its fiery spices. It has its place in the pantry for many hip chefs as the fire is straightforward and the result is a sauce that is thicker than most hot sauces in the United States. It is available at gourmet markets, Asian markets, specialty food stores, and natural food stores.

serves 8

for the paste:

- 1/3 cup Sriracha sauce
- 3 tablespoons fresh lemon juice
- 2 tablespoons honey
- 1 tablespoon garlic, minced

Combine all in a small bowl. Reserve.

for the ribs:

- 2 slabs baby back ribs, about 3 pounds
- Kosher salt and freshly ground pepper to taste
- 2 12-ounce bottles of beer, your favorite
- Spray bottle for the beer

Rinse ribs and pat dry. Loosen the thin papery membrane that runs along the underside of the ribs with a knife and pull it off with your fingers. Sprinkle with salt and pepper and then coat the racks of ribs liberally with the paste by spreading with a rubber spatula. Refrigerate, covered for at least 2 hours or overnight for best flavor.

Preheat the grill to medium.

Place the beer in a spritzer.

Place the ribs, bone side down, on indirect side of the grill and close the lid. Grill, basting gently with the paste and then spritzing with the beer every 10 minutes, until ribs are tender, cooked through, and meat shrinks away from the bones, 40 to 50 minutes total. Make sure to keep the paste fully spread on the ribs while basting.

Serve any remaining paste along with the ribs.

Vietnamese Rice Noodles with Pork or Beef and Vegetables

Bun thit nuong *is a simple and favorite meal in Vietnam consisting of cold rice vermicelli mixed with fresh vegetables and topped with hot barbecued pork or beef. Since I discovered this refreshingly light yet fulfilling meal, I often crave the warm caramelized crunch of the grilled meat atop the fresh, vibrant vegetables and cool rice noodles. Served with* nước chấm *sauce, it is a good summer dish—light, low in fat, healthy, and refreshing.*

Vietnamese recipes use a very diverse range of herbs, including lemongrass, mint, cilantro, and Thai basil leaves. Traditional Vietnamese cuisine is greatly admired for the freshness of its ingredients and for its healthy eating style.

Nước chấm is the all-purpose Vietnamese dipping sauce often served as a condiment.

serves 4-6

for the meat and marinade:

1 pound raw pork loin or beef sirloin steak, partially frozen, then thinly sliced
3 medium scallions, minced
2 medium cloves garlic, minced
½ teaspoon Thai or serrano chiles, minced
2 teaspoons sugar
2 tablespoons fish sauce
Juice of 1 medium lime
Kosher salt and freshly ground pepper to taste

1 8-ounce package rice vermicelli
2 tablespoons peanut or canola oil
3 medium carrots, grated
2 cups mung bean sprouts
2 cups green leaf or romaine lettuce, shredded
1 cup cucumber, peeled, seeded, and thinly sliced
½ cup fresh cilantro sprigs
¼ cup fresh mint leaves
¼ cup roasted peanuts, roughly chopped
1 cup nước chấm

In a large bowl, mix the pork or beef, scallions, garlic, chile peppers, sugar, fish sauce, lime juice, salt, and pepper. Set aside to marinate 15 to 30 minutes. Remove and pat dry.

Place the rice vermicelli in a heatproof bowl and cover with boiling water; let sit for 5 to 8 minutes until softened through. Drain, rinse in cold water, and set aside.

to grill the meat:

Heat the grill to high. Sear the pork or beef until nicely browned but not dry. Remove from the heat and slice very thinly. Keep warm for a moment while building the salad.

for the *Bun thit nuong*:

Add equal amounts of the carrots, sprouts, and lettuce to each serving bowl. Place equal amounts of rice noodles on top of the vegetables. Then place a spoonful of meat on top of the noodles. Place the sliced cucumber over and then lay sprigs of cilantro and mint over the top.

Sprinkle peanuts over each dish and serve with nước chấm or on the side to pour as desired over the noodles.

for the simplest nước chấm:

1 part lemon or lime juice
1 part fish sauce, nước chấm
1 part sugar
2 parts water

Combine ingredients in a small bowl and serve as a sauce for the rice-noodle salad.

Thai-Seared Pork or Poultry on Chop-Chop Herb Salad

Pork, chicken, or turkey tenderloins all marry beautifully with the paste and the lively chopped-herb salad. The vivid blue flowers and leaves of the herb, borage, mirror the flavor of the cucumbers. The dish is an abundance of textures and vibrant flavors . . . plus it's naturally nutritious. Make sure to plan ahead to allow enough time for marinating.

serves 4

for the Thai grilling paste:

makes about 1 cup

1 2-inch-long piece fresh ginger root, peeled and coarsely chopped
2 medium cloves garlic
1 small jalapeño, seeded
3 tablespoons soy sauce
2 tablespoons dark brown sugar
2 teaspoons fish sauce
2 teaspoons freshly squeezed lime juice, more to taste
½ cup chicken broth

4 1-inch-thick pork loin chops or boneless, skinless chicken breast halves or turkey breast

Place the ginger, garlic, jalapeño, soy sauce, brown sugar, fish sauce, and lime juice in a food processor or blender. Puree, scraping down the sides with a spatula. Add some of the chicken broth until the consistency of molasses—thick, but spreadable.

Completely cover the surface of the pork or poultry with the paste. Place on a plate and refrigerate for 4 to 6 hours to marinate.

to grill the meat:

When ready to cook, preheat the grill to medium-high.

Grease the grill rack, place the pork or poultry on the grill, and sear each side to caramelize, about 5 to 8 minutes per side. When the meat is done, transfer to a cutting board and let cool for about 5 minutes. Thinly slice.

for the Chop-Chop Herb Salad:

1 cup fresh basil, coarsely chopped
1 cup fresh cilantro, coarsely chopped
1 cup fresh scallion greens, coarsely chopped
¼ cup borage flowers, if available
1 cup cucumber, peeled and thinly sliced
Asian toasted sesame oil, for drizzling
Fresh lime wedges for garnish
Cooked rice for serving on the side, optional

In a medium bowl, toss the basil, cilantro, scallions, borage flowers, and cucumber. Drizzle with about 1 tablespoon of sesame oil and toss again. Line serving plates or large bowls with the salad and then top with slices of pork or poultry. Serve with lime wedges.

Lamb

Simply Chic Grilled Lamb Sirloin and Pappardelle
Rack of Lamb with Caesar Sauce
Lamb Stuffed with Figs, Chevre, and Port
Lamb Chops with Yogurt and Mint Chimichurri Sauce
Low-and-Slow Grilled Leg of Lamb
Pulled Slow-Grilled Lamb Shoulder
Yogurt Pomegranate Lamb Brochettes on Rosemary Skewers
Peppered Lamb with Herb Hazelnut Sauce
Brined Lamb Shoulder Steaks with Gazpacho Sauce
Yugoslavian Smoked Lamb Shank Kupus
Spinach-and-Feta-Stuffed Leg of Lamb

GRASSFED LAMB: NUTRITION AND FLAVOR

Sheep are ruminants and wonderful grazers, so grassfed lamb offers all the nutritional benefits described earlier in this book. And yet, most lamb that you purchase in markets comes from sheep that have been kept in barns and fed diets of grain and hay . . . not the menu for raising healthy, stress-free animals. Ask about this when you visit farms.

When buying farm-fresh lamb, David Greene, an excellent sheep farmer and quite a chef, gives us a few pointers: animals should be no older than 4 to 6 months when processed, and lamb meat should be hung at the meat shop for only 3 to 7 days. There is not much fat on the lamb and if it hangs any longer than this, the meat will become too dry. David says that his favorite cuts of lamb for grilling include burgers made on the large side—about 1/3 pound for more robust flavor; shanks, usually a lower-cost item, but very lean with excellent flavor; and boneless leg of lamb that's ideal for barbecuing. Mutton comes from a sheep that is more than two years old and is not as popular in the United States as it is in the Middle East. The sheep may be male or female, although mutton from rams can be extremely gamy due to their hormonal balance. The meat is tougher because the animal is older but it also has a more developed flavor. Since mutton is so tough, it needs to be cooked using the low-and-slow, moist-cooking process, meaning slowly, at low temperatures, and with lots of moisture.

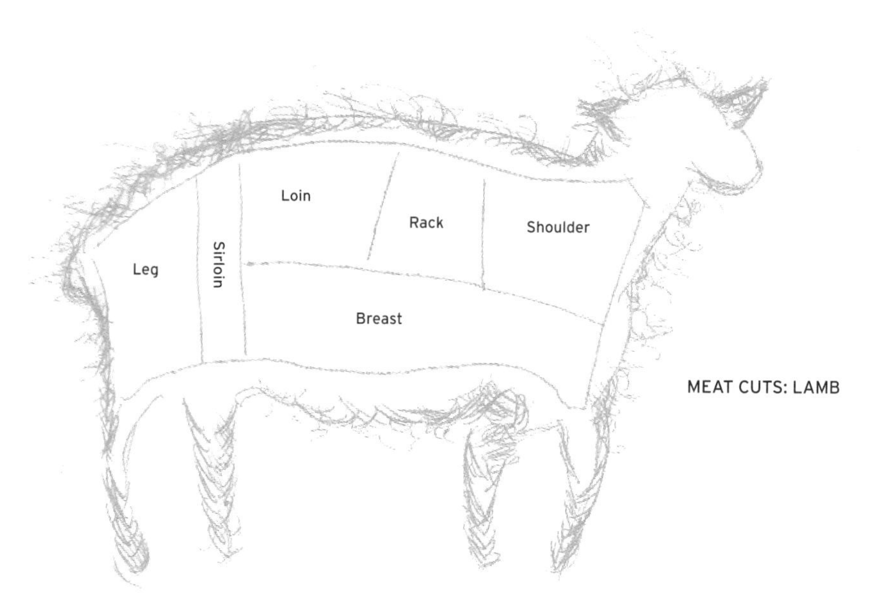

MEAT CUTS: LAMB

Simply Chic Grilled Lamb Sirloin and Pappardelle

I discovered lamb sirloin steak in my cooking school where I created recipes for a variety of cuts of lamb. It is prepped the same way you would prep a basic beef sirloin steak: very simply. The luscious, wide, pappardelle noodle is sturdy enough to sop up all of those vibrant aromatics that serve to embellish the lamb. Oh what a fabulous dinner!

serves 4

for the lamb:

1 pound lamb sirloin steak, at least 1-inch thick
Kosher salt and freshly ground pepper

for the pasta:

8 ounces pappardelle pasta
1 teaspoon chopped fresh thyme, plus extra for garnish
1 medium clove garlic, minced
2 teaspoons grated lemon zest plus juice from 1 lemon
4 tablespoons butter, cut into small pieces
Kosher salt and freshly ground pepper
Lemon wedges, for garnish

Preheat the grill to high heat; set up for direct grilling.

Dry the meat using paper towels. Lightly salt and pepper and let rest on a cutting board for 10 minutes.

Grill over direct heat for 3 to 5 minutes per side until medium rare. Remove from grill and let rest on a cutting board for about 5 minutes, reserving all juices. Slice into ½-inch thick strips.

Cook pasta in a large pot of boiling salted water until al dente, according to package instructions. Reserve ¾ cup cooking water; drain pasta and return to pot.

Add thyme, zest, lemon juice, and reserved cooking water back into the pasta in cooking pot; stir in the butter and add salt and pepper to taste.

Place a large mound of the pasta on a serving platter and pile on top the lamb sirloin strips and drizzle with lamb juices. Garnish with lemon wedges and thyme.

Rack of Lamb with Caesar Sauce

An elegant star in our lamb repertoire certainly must be the rack of lamb from one of our favorite pasture-based farmers, Jamison Farm, which donated some of its prized lamb for our recipe testing. The Caesar sauce is quite thick so add a few drops of water if you prefer a thinner texture.

serves 2

1 rack of lamb, bones "Frenched," and cap removed
1/3 cup port wine
Tip: Remove the thick cap of muscle-streaked fat covering the chops by slicing along the seam and pulling away fat as you go.
Kosher salt and freshly ground pepper to taste

Sprinkle the racks of lamb with port wine. Let season, refrigerated, for at least 1 hour.

for the Caesar sauce:

makes about 1 cup

2 medium cloves garlic, peeled
1 whole raw egg
2 teaspoons Dijon or spicy mustard
1 anchovy fillet
1 teaspoon Worcestershire sauce
2 tablespoons fresh lemon juice
2 tablespoons freshly grated Parmesan cheese
Freshly ground pepper to taste
¾ cup extra-virgin olive oil
Kosher salt to taste

Pasteurizing raw eggs

Sally Fallon states there is no problem eating the raw yolk from pastured eggs except that too many raw whites may cause digestive problems. Just to erase any doubts, pasteurizing is very simple. Place the eggs in a small pot with cold water. Put the water on medium heat and stand by to watch as the temperature rises. You don't want the temperature of the water to exceed 150°F. You can use a thermometer in the water or let the temperature get to 140° to 150°F, the stage before bubbles start to form. When you reach this temperature, maintain it, then lower the heat and keep the eggs in the water another 3 to 5 minutes.

In a food processor or blender, place the garlic, egg, mustard, anchovy, Worcestershire, lemon juice, Parmesan, and pepper. Process the ingredients to mix thoroughly for 1 minute. With the processor on, add the olive oil in a thin stream; when the mixture begins to thicken, you will hear a change in the motor of the processor. Scrape the sauce down from the sides and add salt to taste. You want the sauce to be thick so it will adhere to the lamb as it grills.

Preheat 2/3 of the grill to high or the equivalent of about 450°F. Prepare the lamb by coating with salt and pepper and then spreading with a thin layer of the Caesar sauce. Save the remainder to serve with the plated roast.

Put the rack of lamb on a small, flat pan or heavy aluminum foil and place on the grill near the flame, but not on it. Cover with the lid, and cook about 10 minutes. Flip the lamb over and cook another 5 minutes. Test the center of the meat with a meat thermometer that now should read 135°F for medium-rare. Remove the lamb from the grill, place on a work surface, cover loosely with foil, and let rest for about 7 minutes before carving into chops.

Serve with additional sauce.

Lamb Stuffed with Figs, Chèvre, and Port

I created this recipe when I became enthralled with how well fresh figs paired with olives. Who knew the combo could be so splendid?

Many people are not aware of this, but 100 percent pastured lamb is best eaten medium-rare. The fig-and-olive paste in this recipe is a sort of tapenade that is versatile enough for sandwich spreads, draped over soft cheeses, or topping crostini. Bring the lamb to room temperature an hour before grilling.

serves 8

for the tapenade:

4 large garlic cloves, minced
2 medium shallots, coarsely chopped
2 teaspoons fresh rosemary
1 teaspoon fresh mint
Zest and juice from 1 medium lemon, divided use
½ teaspoon kosher salt
1 cup pitted Kalamata olives
1 cup fresh figs, chopped or ⅔ cup dried and very finely chopped
1 tablespoon extra-virgin olive oil
Kosher salt to taste and freshly ground pepper

for the lamb:

1 4-pound butterflied leg of lamb
Kosher salt and freshly ground pepper to taste
1 cup fresh soft chèvre
¾ cup port wine

for the tapenade:

In a food processor combine the garlic, shallots, rosemary, mint, lemon zest, and salt. Gradually add the olives, figs, and lemon juice. Add the olive oil slowly until all becomes a paste. Taste now to add any salt and pepper.

Place the open lamb, skin side down, on a work surface. Spread firmly with the tapenade and then spread with the soft chevre. Begin rolling from the shortest side, enclosing the filling. Tie the rolled lamb with kitchen string at 1-inch intervals making sure to enclose the filling. You may need small skewers for the ends.

Heat the grill to medium-high.

Sprinkle the lamb with salt and pepper; put over direct heat for about 10 minutes to sear on all sides. Once seared, place the lamb in a roasting pan or cast-iron vessel on the grill and move to indirect heat. Pour the port over the lamb, close the grill lid, and roast 20 minutes, basting twice during the cooking process with pan juices. When the lamb reaches 135°F on a meat thermometer, transfer the roast to a cutting board and let rest 10 minutes, loosely covered with foil.

Tilt the pan and skim excess fat from the top. Place remaining juices in a serving container.

Remove the strings and slice the lamb crosswise into ½-inch-thick slices. Arrange on a platter, drizzle the juices over the slices, and serve with additional pan juices on the side.

Lamb Chops with Yogurt and Mint Chimichurri Sauce

The magazine group Edible Communities Publications (www.ediblecommunities.com) is a treasure trove of sustainable and newsy articles and recipes. This recipe from their edible WOW, *Southeastern Michigan edition, is adapted from the book* Romance Begins in the Kitchen *by Dawn Bause, Modesta DeVita, and Nidal Daher (BauseHouse Publication, 1998). The chimichurri is by Pam Aughe, editor of* edible WOW. *Spectacular it is!*

serves 4

for the lamb chops:

2 tablespoons olive oil
1 medium clove garlic, crushed
1 tablespoon balsamic vinegar
½ teaspoon salt
Freshly ground pepper
4 small lamb chops, loin cut, medium-thick

Combine olive oil, garlic, vinegar, salt, and pepper in a small bowl. Brush mixture on both sides of the lamb chops. Place chops on a plate, cover and marinate 1 to 2 hours in the refrigerator.

Behind the lamb renaissance

Over the past 30 years, high-school sweethearts John and Sukey Jamison have produced some of the country's best-tasting lamb. A very talented cook herself, Sukey has created a line of hand-prepared lamb dishes that include her award-winning lamb pie and a lamb stew developed in collaboration with the late, great Jean-Louis Palladin.

for the sauce:

2 cups loosely packed fresh mint
2 cups loosely packed fresh parsley
2 medium cloves garlic
¼ cup extra-virgin olive oil
1/3 cup Greek-style yogurt
¼ teaspoon freshly ground pepper
Pinch of kosher salt
Juice of 1 whole lime
2 tablespoons water

Combine all ingredients in a blender. Mix until well combined, stopping to scrape down the sides as needed. Set aside.

Preheat grill to medium-high.

Cook lamb on grill about 2 to 4 minutes per side, depending on the thickness of the chops or until medium-rare. Serve with chimichurri sauce.

Low-and-Slow Grilled Leg of Lamb

The savvy cooks of South America often use low heat when grilling meat—no tall orange flames, only glowing embers. This technique allows lamb to cook slowly and remain juicy until it just falls apart. Our recipe is inspired by the Georgia Slow Food chapter—a regional arm of Slow Food USA (www.slowfood.com)—which advocates the same philosophy of allowing plenty of time for the art of cooking.

serves 8 to 10

1 5- to 6-pound whole leg of lamb, bone in
3 medium garlic cloves
3 tablespoons Dijon or spicy mustard
2 tablespoons extra-virgin olive oil
Juice of 1 large lemon
1 tablespoon fresh rosemary leaves
1 teaspoon kosher salt
Freshly ground pepper

Discard the meat's tough membrane and excess fat; place the lamb in a large grill-safe baking dish. First, purée the garlic in a food processor or blender, add the remaining ingredients, blend to a paste, and then spread over the lamb. Cover and marinate in the refrigerator 2 to 8 hours.

Preheat the grill to high.

After 15 minutes, turn off the center burners and reduce the outer burners to medium.

Place the lamb in the baking dish on the indirect-heat part of the grill with the lid down, 1¼–1¾ hours. If using charcoal, add 10 briquettes after 1 hour to maintain heat. The lamb is done when a meat thermometer registers 140°F inserted into the thickest part of the roast (avoid touching the bone). Let rest at room temperature for 15 minutes before carving.

Pulled Slow-Grilled Lamb Shoulder

At Jamison's Farm in Latrobe, Pennsylvania, half of the husband-and-wife team was educating people on the lesser-known cuts of lamb we've come to love. Sukey Jamison decided to slow roast the lamb shoulder with some onions thrown in for additional flavor. Then she started to sell the pulled meat and onion sandwiches every Saturday, from June through October, at the farmers market in Latrobe where she sells her cuts of Jamison lamb. Now Sukey has created a following that begs for those melt-in-your-mouth, grill-seasoned, pulled-lamb sandwiches. Sukey says that pulled lamb works in recipes such as shepherd's pie or as filling for ravioli.

serves 4 to 6

1 whole lamb shoulder, approximately 5 to 7 pounds
Kosher salt and freshly ground pepper to taste
1 whole bulb of garlic, top ¼ cut off to expose the cloves
6 stalks of thick celery, coarsely chopped
3 sweet onions, halved crosswise

Preheat half the grill to medium-high.

Season the entire shoulder with salt and pepper. Place directly on the grill over medium-high heat to sear. Grill each side about 4 minutes until browned, but not charred.

Place the vegetables and garlic on a double layer of heavy foil. Using tongs, place the seared meat on top of the veggies. Wrap all with foil—be sure to enclose all edges of the lamb—and move the package to the indirect heat side of the grill. Close the lid to retain an even heat and cook slowly 3 to 4 hours. Make sure to check at least every hour until the lamb pulls easily from its bones when done.

Remove the package from the grill and place on a flat workspace. Pull the lamb into shreds, chop the onions, gently squeeze the cooked garlic from its skins, and toss all together.

Serve as pulled lamb and veggies on a roll as a sandwich, or with mashed potatoes using the drippings in the foil as sauce. The lamb can be grilled a day ahead, but pulling lamb is easier when lamb is still warm.

Yogurt Pomegranate Lamb Brochettes on Rosemary Skewers

We very successfully tested and served this lamb creation at the Cook Local: Save the Bay *class I gave at the Chesapeake Bay Foundation's Philip Merrill Center in Annapolis. The class was given a "perfect" rating as our local and sustainable products were paired with fine local wines . . . all in the gorgeous fall setting of the Chesapeake Bay.*

serves 4 to 6

- 1 cup plain yogurt
- Kosher salt and freshly ground pepper to taste
- 1/3 cup pomegranate vinegar
- 2 medium cloves garlic, coarsely chopped
- 1 tablespoon fresh rosemary, minced
- 2 pounds boneless lamb shoulder, cut into chunks
- Sturdy rosemary branches, soaked at least 30 minutes in cool water
- Extra-virgin olive oil, about 1/3 cup

In a nonreactive bowl, combine the yogurt, salt, pepper, vinegar, garlic, and minced rosemary. Add the lamb cubes and toss to coat. Marinate at least 4 hours or overnight.

When ready to grill, remove the lamb from the marinade and pat dry.

Start a charcoal or wood fire or heat a gas grill; grill should be moderately hot. Thread lamb onto rosemary branches, 3 or 4 chunks of lamb per rosemary skewer. Brush lightly with olive oil just before grilling.

Grill, turning skewers as each side browns, taking care to avoid flare-ups; total cooking time should be from 6 to 10 minutes for medium-rare. Meat continues cooking a bit more after you remove it from the grill, so keep this in mind.

Peppered Lamb with Herb Hazelnut Sauce

Yes, this dish is lush and rich and you know you deserve the splurge. Do try and use local and sustainably raised cream (raw would be the ultimate) to go with your locally raised lamb.

serves 4

½ cup whipping cream
⅔ cup chicken stock
3 young fresh carrots (about 3 inches long), julienned
2 teaspoons fresh lemon thyme, finely chopped
3 tablespoons freshly snipped garlic chives
Kosher salt
4 tablespoons coarsely cracked pepper
4 lamb shoulder chops, preferably boneless, about 4 ounces each, trimmed
½ cup hazelnuts, toasted and coarsely chopped

for the sauce:

Heat the cream, chicken stock, carrots, and lemon thyme in a heavy, small saucepan over medium-high heat. Stir occasionally, until the carrots are tender, yet still crunchy. To preserve chives, stir in now. Season with salt and pepper; cover and keep warm on very low heat or make sauce up to one day in advance and reheat gently.

for the lamb:

Heat the grill to high.

Press pepper onto both sides of lamb and then sprinkle completely with salt. Grill lamb about 4 minutes per side for medium-rare or 130° to 135°F on a meat thermometer. Transfer lamb to plates.

Combine most of the hazelnuts and sauce and spoon over the lamb; sprinkle remaining nuts over the sauce as a garnish and add a few more chives.

Brined Lamb Shoulder Steaks with Gazpacho Sauce

Another recipe that makes great use of a lesser-known and infrequently used cut of lamb. Buttermilk, a perfect natural tenderizer, acts here like brine. The gazpacho sauce can be made up to 2 days in advance and gets better as it sits.

serves 4

4 lamb shoulder steaks, about 1-inch thick
2 cups buttermilk, from your farmer or farmers market, if possible
1 teaspoon kosher salt
1 small tomato, finely diced, for garnish
¼ English cucumber, peeled, finely diced, for garnish

for the sauce:

¾ pound fresh tomatoes, peeled and seeded, plus additional for garnish
¼ red onion
2 medium cloves garlic, peeled
3 ounces English cucumber, about ⅓ of a cucumber, peeled, plus additional for garnish
2 tablespoons salt-free tomato juice
1 tablespoon grated lemon zest
2 teaspoons fresh lemon juice
2 to 3 tablespoons sherry vinegar
¼ cup extra-virgin olive oil
Kosher salt and freshly ground pepper, to taste

to brine the lamb:

Place the lamb, buttermilk, and salt in a medium-sized bowl and stir gently. Cover and refrigerate overnight.

"Fall-off-the-bone meat"

The first rule of shredding cooked meat is that the meat or poultry be so well done it easily falls from the bone. "Fall-off-the-bone meat" is the title we gave to our grandmother's juicy beef roasts. While working on a cutting board, take a fork in each hand and literally pull the meat apart. In other words, tear it into shreds.

---

to make the sauce:

Place the tomatoes, onion, garlic, cucumber, tomato juice, lemon zest, lemon juice, and 2 tablespoons of the vinegar in a blender; pulse a few times to combine. With the motor running, add the olive oil in a steady stream and blend until well mixed, but still slightly chunky. Season to taste with salt and pepper, and add more vinegar if necessary.

for the steaks:

When ready to grill, remove the lamb from the brine, pat dry.

Heat the grill to medium-high.

If the steaks are about 1-inch thick, place directly over the flame and grill each side for about 6 minutes or until a meat thermometer reads 130° to 135°F for medium-rare.

to serve:

Spread the gazpacho sauce thickly all over the steaks and garnish with the diced tomato and cucumber.

Yugoslavian Smoked Lamb Shank Kupus

This Serbian smoked lamb and kale stew is a complex play of meat and vegetable flavors that comes straight from Yugoslavia via my remarkable cooking experiences in California amid a group of Yugoslavian topnotch produce growers. We had an amazing smokehouse, Corralitas Market, where we would buy smoked lamb breasts for this Old World dish . . . majoring in kale and, of course, lamb.

serves 8 with lots of leftovers

to smoke the lamb shanks:

4 meaty lamb shanks
4 cups hardwood chips, soaked in water for 30 minutes

Place the lamb shanks on a piece of heavy foil and place directly over the flame to brown. Let grill 10 minutes, and then, with tongs, turn to brown the other side. Move the shanks on foil to the indirect heat of the grill. Place about ¼ of the soaked chips on another piece of foil punctured with holes in a few places to allow smoke to escape; put this packet of soaked chips directly on a flame to create smoke. Replace the wood chips as needed. Smoke the lamb shanks about 3 hours or until deep brown and meat pulls away from the bone.

for the stew:

2 bay leaves
2 6-inch sprigs fresh rosemary
4 to 5 peppercorns, crushed
12 medium-sized root vegetables, such as turnips, carrots, and parsnips, mixed
3 large sweet potatoes, quartered
2 onions, peeled and quartered
1 pound small, locally grown potatoes
6 cups chopped kale, Savoy cabbage, or Swiss chard, or a mixture of greens
4 medium cloves garlic, minced
Kosher salt and freshly ground pepper
1 sprig fresh rosemary, finely chopped

Place the smoked meat, bay leaves, rosemary sprigs, and peppercorns in a stockpot; cover with 6 quarts of cool water and bring to a boil. Reduce heat, and simmer, covered, for close to 3 hours or until meat easily falls from the bone. Skim most of the fat from the top of the soup. A small amount of fat is fine because this pastured-lamb fat has health benefits.

Add the root vegetables, sweet potatoes, onions, potatoes, and cabbage. Cook another 45 minutes or until all vegetables are tender.

Just before serving, remove the bay leaves; add the garlic, salt, pepper, and about 2 tablespoons chopped rosemary.

Spinach-and-Feta-Stuffed Leg of Lamb

Yes, I know most of us associate lamb with garlic and rosemary. When I first visited Greece, however, I became enamored with the flavors of the lamb I consumed with great enthusiasm on the Greek Islands. It wasn't so much the quality of the lamb that caught my attention but the olive oil, garlic, lemon, spinach . . . on and on. This recipe is truly scrumptious, especially with the fabulous pastured lamb you pick up from the farm.

serves 8

for the marinade:

1 cup dry red wine
Zest and juice from 1 medium lemon
½ cup extra virgin olive oil
2 tablespoons Dijon or spicy mustard
1 tablespoon fresh basil, finely chopped
4-pound leg of lamb, boned and butterflied

Combine all marinade ingredients in a 9 × 13-inch glass baking pan. Open the lamb and add to the marinade, turning to coat. Wrap the dish with plastic and marinate, refrigerated, up to 6 hours or overnight.

for the stuffing:

1 pound fresh baby spinach, lightly wilted, squeezed dry, and coarsely chopped
1 pound ricotta cheese
6 medium scallions, finely chopped
1 egg, beaten
1 cup pine nuts, toasted
1 teaspoon dried oregano leaf, crushed
Kosher salt and freshly ground pepper to taste

Place all ingredients in a medium bowl and stir with a spoon to mix well. Refrigerate until the lamb is ready to prepare for grilling.

for the lamb coating:

½ cup Dijon or spicy mustard
1 tablespoon honey
2 medium cloves garlic, finely chopped

Remove the lamb from the marinade; blot dry. Lay it flat on a work surface and spread the stuffing over the inside of the lamb. Work on a slight diagonal and roll the lamb to about 5 inches in diameter. Cut off any sinewy flap of skin. Tie the rolled lamb with butcher's string at 1-inch intervals making sure to tuck in the filling.

At this time, mix the ingredients for the lamb coating.

Heat the grill to medium-high.

Place the lamb over direct heat for about 10 minutes to sear. Then move the lamb to a roasting pan or cast-iron vessel for the grill on indirect heat. Spread the mustard coating over the entire outside surface of the rolled lamb. Close the grill lid and roast the meat for 45 minutes, turning 2 times. When the lamb reaches a temperature of 140°F on a meat thermometer, transfer the stuffed lamb to a cutting board and let rest 15 minutes.

Pour any pan drippings into a measuring cup and allow the fat to float to the surface; skim and discard. Remove the strings. Cut the lamb crosswise into ½-inch-thick slices. Arrange on a platter, spoon the pan drippings over the lamb, and serve.

Turkey and Chicken

Turkey Breast with Rita's Lacquer Sauce
The Whole Holiday Bird on the Grill: Brined Heritage Turkey with Chunky Cherry Glaze
Good Ol' Barbecued Turkey Breast
River Dinner Meteorite Turkey
Indian-Style Turkey Skewers on Spaghetti Squash
Turkey-in-the-Hole
Turkey Grill with Lemon-Herb Yogurt Sauce
Hoisin Citrus-Tea Smoked Chicken
Beer-Can Pulled Chicken with Smoked-Tomato Salsa
Chicken Breasts 101: Grilled with Double-Fig Glaze
Aromatic Buttermilk-Basted Chicken
Brined Lime-Pineapple Chicken
Grill-Roasted Pear-and-Bacon Stuffed Chicken
Sweet and Spicy Glazed Chicken
Soft Shredded-Chicken Tacos
Spanish Stuffed Chicken Packets with Romesco Sauce
Chicken Salad Niçoise
Za'atar Chicken Wraps
Moroccan Chicken
Pasture and Garden-Vegetable Kebabs
Poultry Stock

TURKEYS AND CHICKENS THAT TASTE LIKE TURKEYS AND CHICKENS

Raising chickens on pasture is quite a contrast to the very industrialized model of raising chickens in huge groups of up 30,000 birds crammed into a poultry house. The difference in flavor and texture is an equal contrast and one not to be missed. Robin Way, a Maryland-based grass farmer, tells of a customer who said: "Finally, a chicken that tastes like chicken. I don't even have to flavor it. It has flavor right from the get-go. And it's not mushy like the chicken I get at the grocery store." Most pastured chickens are raised either in low chicken pens with open bottoms that are rolled to fresh grass daily (often called chicken tractors) or in pastured areas protected by a woven net of electric fencing, more to keep predators out than to keep the chickens in. Do not be fooled into thinking that every product labeled *free range* automatically means it was pasture-raised. Free range is an ambiguous term used as a marketing tool that can be applied legally to chickens with access to an outside area they may or may not ever enter, and that may or may not have a blade of grass. Check with your farmers to see that the birds are truly pasture-raised; that is, they can get as much fresh grass as they want to eat. Pastures provide a clean, dust-free environment for the birds that is far healthier for them than the ammonia-choked and dusty poultry houses where most chickens are raised.

Chickens cannot live solely on grass because, unlike ruminants, they will not get sufficient nutrition from eating leafy greens alone. Although they do pick up some nutrition from eating plants in the pasture, they need the bulk of their diet supplemented with grains grown in crops fields. This means that the meat of pasture-raised chickens is going to be higher in omega-3 fatty acids than conventionally grown birds; to be higher in omega-3s the chickens must receive a substantial amount of omega-3 feed, such as flaxseed, which contains three times as many omega-3 as omega-6 fatty acids. Pastured chicken meat has a richer texture, and a robust flavor . . . from the get-go.

Turkey Breast with Rita's Lacquer Sauce

I created my Lacquer Sauce (page 201) when searching for a universal glaze with outstanding flavor and minimal ingredients. Thus this sweet, salty, spicy sauce or glaze was born. It lends an exotic character to whatever it graces, be it poultry, pork, or even seafood. Any extra glaze can be stored in the refrigerator indefinitely.

serves 6 to 8

- 4 tablespoons toasted sesame oil
- 2/3 cup apricot preserves
- 4 tablespoons low-sodium soy sauce
- 3 medium cloves fresh garlic, minced
- 1 2- to 3-pound pastured turkey breast

In a small bowl, mix all ingredients; use part of the glaze to marinate and the remainder to brush on the meat while grilling.

Place the turkey breast in a large zipper storage bag, pour the marinade over it, and marinate chilled at least 2 hours or as long as overnight.

Preheat half the grill to medium-high.

Remove the turkey from the marinade and place breast-side down on a large sheet of greased heavy foil. Pour the remaining marinade into a small saucepan to create the glaze, heat just to the boiling point, and cook about 2 minutes.

Add the turkey breast to the hot portion of the grill and grill about 15 minutes to brown. Flip the turkey over, still on the foil, and grill another 15 minutes. Move the turkey on the foil to the indirect side of the grill and baste with more glaze. Continue cooking for another 35 to 50 minutes, basting every 15 minutes, until the internal temperature on a meat thermometer reaches 160° to 165°F.

Let rest 5 to 10 minutes, tented loosely with foil, and then carve.

The Whole Holiday Bird on the Grill: Brined Heritage Turkey with Chunky Cherry Glaze

Make sure to bring the bird to room temperature before cooking. Most heritage turkeys are sold fresh. The general rule allows 10 minutes per pound roasting time for heritage or free-range turkeys; we recommend filling the cavity with lots of fresh herbs of your choice.

Use a brining bag or double up with two heavy-duty, unscented trash bags and, if the bag won't fit in the fridge, put it in a cooler or ice chest large enough to hold the turkey. If your holiday bird is small, like the one we tested, it should fit in your refrigerator.

serves 8 to 10

1 fresh heritage turkey, about 12 pounds

for the brine:

2/3 cup kosher salt
1 cup light brown sugar
1 tablespoon freshly ground pepper
3 dried bay leaves, broken into pieces
6 medium sprigs fresh thyme
2 gallons cold water

Remove the giblet bag from the turkey along with any extra internal fat and the small, fine, pin feathers. Rinse the bird well under cold water. If using a large, sturdy bag for brining, combine the salt, sugar, pepper, bay leaves, and thyme in the bag; add the cold water. Stir until the sugar and salt dissolve. Add the turkey; there should be enough liquid to completely cover it. Press the air out of the bag and close tightly. Keep the turkey cold with bags of ice, which will also help keep it submerged in the brine. Brine 12 to 24 hours.

Alternatively, place turkey and brine in a large pan or bowl. Refrigerate for 12 to 24 hours. If turkey floats to the top, weigh it down with a plate and cans to keep it submerged.

for the glaze and sauce:

makes 2 1/4 cups

- 1½ cups McCutcheon's Cherry Salsa
- ¾ cup Boar's Head Sugar-and-Spice Glaze
- ½ cup dried cherries, for the sauce
- ¼ cup balsamic vinegar

Combine the salsa and glaze. Remove 1 cup of the glaze mixture for the serving sauce.

Place the dried cherries in a small bowl and cover with balsamic vinegar. Let stand at room temperature for 1 hour to plump and then mix well with the reserved cherry glaze.

This is a big PS: The glaze is so good, you will devour it by the spoonful so consider doubling the recipe. Make one and a half times the recipe for a large turkey.

to grill-roast:

Preheat grill to medium-high. Sprinkle pepper in the cavity and rub into the skin. Tuck wing tips under, loosely truss legs, and place turkey on a V-shaped rack in a roasting pan. You may want a layer of heavy foil wrapped around the bottom of the pan to protect it from the flames or use a heavy, disposable roasting pan. Tent breast loosely with foil.

For the first hour, cover the bird with heavy foil to seal, and cook directly over the flame. For the second hour, move the turkey in the roasting pan to the indirect heat side of the grill, remove foil covering and spoon about 1/3 cup glaze over the turkey. Repeat glazing at least 2 more times during the grill roasting.

For doneness, test with an accurate thermometer, making sure not to touch the bone with the tip of the thermometer. The turkey should have an internal thigh temperature of 160° to 165°F.

Let the roasted bird rest 10 to 15 minutes before carving. Serve extra cherry glaze alongside.

Good Ol' Barbecued Turkey Breast

The subtle, smoky flavor of barbecued turkey breast will prove that turkeys are a year-round treat. I have been grilling and smoking turkey—whole and parts—for almost 25 years. These days, whole turkeys and turkey parts are easily found throughout the year.

In this recipe we are using Jim Tabb's barbecue rub because it is the best we have tried. The play of spices is especially brilliant, and it works magic during the slow barbecue process.

This barbecued turkey breast is perfect with some grilled sweet potato wedges and a picnic-style garden salad.

serves 6 to 8

1 4-pound turkey breast
Vegetable oil for greasing the pan

for the barbecue rub:

makes 3 cups

1¼ cups firmly packed dark brown sugar
1/3 cup kosher salt
¼ cup granulated garlic
¼ cup paprika
1 tablespoon chile powder
1 tablespoon freshly ground red pepper
1 tablespoon freshly ground cumin
1 tablespoon lemon pepper
1 tablespoon onion powder
2 teaspoons dry mustard
2 teaspoons freshly ground pepper
1 teaspoon freshly ground cinnamon

Combine all ingredients for the rub; spread under the skin and pat liberally all over the turkey breast. Let rest, covered with plastic wrap, in the refrigerator overnight.

Heat the left and right sides of the grill to medium-high leaving 1/3 of the grill for indirect heat.

Place the turkey, breast-side down, in a lightly oiled cast-iron pan ready for the grill. Set the pan with turkey on a rack over the unlit portion of grill. Close the grill lid and grill-roast for about 1½ hours (depending on size), flipping the meat over halfway during the cooking time, until a meat thermometer registers 160° to 165°F.

Transfer the turkey to a platter and tent loosely with foil. Let rest 10 minutes, saving all the juices.

Slice the turkey and place back on the serving platter, drizzle with the pan juices.

River Dinner Meteorite Turkey

I have come to adore an old Tuscan method of grilling chickens that uses bricks to weigh down a butterflied chicken, resulting in quick and even cooking, and crispy skin. I started organizing what I call "river dinners" on the Severn River in Maryland a while ago. At one of those lively dinners, a few of us went further and tested a small turkey. Not having any bricks available, we borrowed (and later returned) some large rocks from the jetty and covered them with heavy foil to compress the bird, thus enabling it to cook evenly. One clever young fellow named the rocks "meteorites—direct from the sky to our grill."

It's easy to butterfly the turkey by cutting along the backbone with a sharp, serrated knife and then spreading it out flat in a large pan to marinate.

serves 8 to 10

1 cup fresh orange juice
1/3 cup fresh lime juice
¼ cup fresh lemon juice
2 tablespoons extra-virgin olive oil
1 tablespoon fresh oregano, finely chopped
1 tablespoon fresh basil, finely chopped
3 teaspoons salt, divided use
3 medium garlic cloves, minced
1 small pastured turkey, about 12 pounds, neck and giblets removed, and butterflied
1 teaspoon Spanish smoked paprika
1 teaspoon freshly ground pepper
Nonstick cooking spray
1½ oranges with skin
4 foil-wrapped bricks or 2 foil-wrapped large flat rocks for weights

Whisk the juices, olive oil, oregano, basil, 1 teaspoon of the salt, and garlic. Put the turkey in a glass baking dish and pour the mixture over the turkey. Turn to coat; chill for 2 hours or overnight, turning occasionally.

Mix the remaining 2 teaspoons salt, paprika, and pepper in a small bowl.

Heat the grill to medium.

Treat the grill rack with nonstick cooking spray. Slice ½ an orange into ¼- to ⅛-inch-thick slices. Remove the turkey from the marinade and pat dry. Loosen the skin from the turkey breast and slide 2 orange slices between skin and breast; loosen skin from the thighs and slide 1 to 2 orange slices between skin and thighs.

Rub the paprika mixture all over the turkey. Place, skin-side down, on the grill over direct flame. Place the foil-wrapped bricks or stones on top of the turkey to evenly compress it (if using bricks, position 1 brick over the top half of turkey, and 1 brick over the bottom half).

Cover and grill until the skin is crispy and brown, about 15 minutes. Remove bricks or stones. Using 2 large spatulas, flip the turkey. Replace bricks or rocks and cook, covered, for 15 minutes more to brown. Now move the flattened turkey, with the weights, to the indirect portion of the grill and close the lid. Continue grilling about 1½ hours or until the bird is cooked through and registers 160° to 165°F in the thighs and breasts on a meat thermometer. Remove the turkey to a large cutting board and let rest 10 minutes, tented loosely with foil. Carve the turkey as desired.

Indian-Style Turkey Skewers on Spaghetti Squash

Kebabs are popular as a quick and efficient way to grill. We can learn a lot about them from other cultures. Garam masala, for example, is an Indian spice blend, a family favorite among many Indians, that is often made fresh daily and used regularly on kebabs.

In this dish, the strands of spaghetti squash mimic pasta, but with added vitamins and beta-carotene (Vitamin A).

serves 4

1½ cups plain low-fat yogurt, divided use
2 teaspoons garam masala
1 teaspoon Madras curry powder
2 medium garlic cloves, minced
1 tablespoon plus 1 teaspoon salt, divided
½ teaspoon freshly ground pepper
1½ pounds turkey breast tenderloins, cut into 1½-inch-wide × 3-inch-long strips
1/3 cup crumbled feta cheese
3 tablespoons red onion, minced
1 teaspoon fresh lemon zest, finely shredded
2 tablespoons fresh mint, finely chopped, divided use
2 medium red and yellow bell peppers, seeded and sliced into 1½-inch pieces
4 8-inch metal skewers
1 medium baked spaghetti squash, seeded and raked into strands

for the marinade:

Mix 1 cup of the yogurt with the garam masala, curry powder, garlic, 1 teaspoon of the salt, and pepper in a large zipper bag. Add the turkey and marinate 2 hours or overnight.

for the sauce:

In a small bowl, stir together the remaining ½ cup yogurt, the feta, onion, lemon zest, and half the chopped mint; set aside.

Preheat the grill to medium-high.

Thread the turkey alternately with the red and yellow bell peppers onto the metal skewers and discard marinade. Grill the kebabs, turning once, until the turkey is browned and cooked through, about 10 minutes.

Pile the shredded spaghetti squash on a platter, sprinkle with the remaining mint. Remove the turkey from the kebabs and arrange the turkey around the squash. Serve with yogurt-feta sauce.

Turkey-in-the-Hole

Although a Mennonite farmer, Myron Martin, gave us this recipe, it is not specific to the Mennonite community; rather, it is a fun activity that can be done with lots of neighborhood children. It's also quite dramatic so make sure you leave plenty of time—the fire pit itself takes only a mere 5 to 8 hours to be ready for grilling.

The whole group gets involved because the fire pit must be fanned and maintained until it is time to drop the turkey down the hole. I've even heard tales of some groups that have made huge pits and filled them with about 20 turkeys. Both this recipe and the size of the pit are estimated and adjusted for just one bird. In general, the recipe derives from local folklore, but a few of the specifics are our very own.

serves 15

for the brine:

1 15- to 20-pound pastured turkey
Bourbon and Coffee Brine (page 209), or Rita's favorite turkey brine (page 124), Cherry Cabernet Sauce (page 203)
4 ounces, 1 stick, unsalted butter, at room temperature

Remove the giblet bag from the turkey along with any extra internal fat and remove the small, fine, pin feathers. Rinse the turkey, inside and out, under cold water. If using a large, sturdy, plastic bag, combine ingredients in the bag, add the cold water, and stir until sugar and salt dissolve. There should be enough liquid to completely cover the bird. Press all air out of the bag and then close tightly. Keep turkey cold with bags of ice, which will help keep it submerged in the brine. Brine 12 to 24 hours.

If you prefer, place the turkey and brine to cover in a large, deep pan or bowl. Refrigerate for 12 to 24 hours. If turkey floats to top, weigh it down with a plate and cans to keep it submerged.

for the fire pit:

- Heavy-duty aluminum foil
- 20-pound bag mesquite or other wood, if not using wood gathered at the site
- Leafy tree branches for fanning the fire
- Chicken wire to wrap around the turkey
- Meat thermometer

to prepare the pit:

For a 15- to 20-pound turkey, dig a cube-shaped a hole in the ground that measures 24 × 24 × 24 inches. Lay green sapling branches (young trees), 3 inches in diameter, across the hole, completely covering the hole. Build a roaring fire on the saplings. Use enough wood so you can completely fill the hole with red-hot coals once the fire matures. Fanning the fire helps it burn faster.

When the saplings burn through and the fire collapses into the hole, you are ready for your turkey. Figure about 5 to 8 hours to leave plenty of time for your final turkey prep.

Fill the inside of the turkey with ice, which will keep the turkey moist, and rub a good amount of butter on the outside to coat the bird. Wrap the turkey in 4 layers of heavy foil and then wrap chicken wire around the turkey, leaving about 2 feet of wire sticking up. This will serve as your handle to lower and lift the turkey to and from the coals.

Using a shovel, dig a hole in the coals big enough to contain the turkey and drop it into the pit. Completely cover the turkey with coals with at least 2 to 3 inches on its top, and then cover with the earth you dug up. Pack lightly and make sure you keep the chicken wire showing above the earth. You will see the ground smoking while cooking your turkey.

Our turkey was on the small side so it took just 3 hours to cook, but it is worth checking at this point unless you have a whopping big turkey. When checking the bird, pull it up partially and insert a meat thermometer into a few different spots to make sure it is cooked throughout. The thermometer should read 160° to 165°F.

Haul it out, unwrap your masterpiece, and carve as you would any basic bird. Take a bow for a Herculean effort. Well done!

Turkey Grill with Lemon-Herb Yogurt Sauce

If you can find turkey tenderloins from a local farmer, go for it! These pieces are so versatile and, of course, so tender. Keep the tenderloins as thick as possible to retain moistness when grilling. You may want to make the sauce first and have it ready so you can fully concentrate on the grilling.

serves 4

for the sauce:

½ cup plain Greek-style yogurt
3 tablespoons extra-virgin olive oil
Zest and juice of ½ lemon plus more to taste
1 teaspoon garlic, minced
1 tablespoon fresh chives, minced
2 teaspoons fresh dill, minced
1 tablespoon fresh basil, minced
Kosher salt and freshly ground pepper to taste

In a small bowl, whisk together the yogurt, oil, lemon juice, garlic, and herbs. Season to taste with salt, pepper, and more lemon juice if needed. Set aside.

to grill:

Preheat half the grill to medium-high.

Pat the tenderloins dry and then coat with a little extra olive oil; sprinkle with salt and pepper.

Place the tenderloins on the grill over the flame and sear for about 6 to 8 minutes. Turn and brown the other side for another 6 to 8 minutes until the tenderloins are golden brown. Make sure they don't dry out.

to serve:

Lemon slices, for garnish

More fresh chopped herbs for garnish

Place the tenderloins on a serving platter and top with the lemon slices and the additional fresh herbs. Serve the sauce on the side.

Hoisin Citrus-Tea Smoked Chicken

We adore the spicy star anise in Hoisin sauce, which gives it that elusive yet captivating character. We've added piquant fresh lime juice and extra garlic for zip. The smoky-tea grilling method adds a dimension that tastes like lots of work. We fooled them!

serves 2

¼ cup Hoisin sauce
Juice of 1 fresh lime
2 medium cloves garlic, minced
2 teaspoons lemon verbana, finely chopped
2 boneless, skinless chicken breasts

Combine the Hoisin sauce, lime juice, garlic, and lemon verbana in a medium bowl or zipper bag. Add the chicken and marinate 2 hours to overnight.

for the smoking fuel:

¼ cup jasmine green tea or any black tea
¼ cup jasmine rice
¼ cup sugar

When ready to smoke the chicken, heat the grill to high.

For the smoking fuel, combine the tea, rice, and sugar. Spread the mixture evenly in the center of a sheet of heavy aluminum foil. Place a small rack over the tea mixture or form a "snake" out of aluminum foil to put around the tea to support the chicken.

Place the smoking mixture on the foil over the heat of the grill. When the mix begins to smoke, place the chicken, over the tea, on the rack, or balanced on the foil snake. Cover with the grill lid (if you also wrap the lid in foil you will have no clean up whatsoever other than recycling the foil). Smoke for close to 20 minutes and check for doneness.

Serve immediately or cool and use for a great stir-fry or Asian-style chicken salad. The smokiness is enhanced by refrigerating the smoked chicken overnight.

Beer-Can Pulled Chicken with Smoked-Tomato Salsa

This fun and fabulous dish comes in many shapes and sizes. The beer can acts as an aromatic roaster that seasons the bird at the same time that it props it up. The angle (literally) for grilling a whole chicken in this recipe yields superb moist meat. We've added our irresistible smoked-tomato salsa on the side because it always gets rave reviews. Two winners here!

serves 4 to 6

1 4-pound whole chicken, fat and giblets removed, rinsed and patted dry
2 tablespoons extra-virgin olive or vegetable oil
Kosher salt and freshly ground pepper
3 tablespoons of your favorite dry spice rub
1 can beer, not a glass bottle because of the possibility of breakage

Rub the chicken lightly with oil, and then rub inside and out with salt, pepper, and the dry rub. Set aside.

Open the beer can and take a few gulps (leave the can half-full). Place the beer can vertically in a grill-safe roasting pan; place the bird cavity firmly over the can, and transfer the pan—with the beer can and bird—to your grill, centered on the rack. Be sure the bird remains upright.

Grill the chicken over medium-high indirect heat, with the grill cover on, for approximately 1 to 1¼ hours or until the internal temperature registers 160° to 165°F in the thigh area or until the thigh juice runs clear when pierced with a sharp knife. Remove from the grill and let rest 10 minutes.

When cool enough to handle, pull the meat from the chicken, shred into bite-sized pieces, and place on a serving platter. Place a bowl of the salsa on the same platter, serve, and stand back to avoid the rush.

for the salsa:

makes about 2½ cups

1 pound-5-medium vine-ripened tomatoes, red, orange, or both
½ pound smoked tomatoes (page 212)
1 fresh serrano or jalapeño pepper
¼ medium onion, preferably white
½ cup fresh finely chopped cilantro sprigs, optional
1 teaspoon minced garlic
1½ tablespoons fresh lime juice
Kosher salt and freshly ground pepper to taste

Quarter and seed tomatoes. Cut tomatoes into ¼-inch dice and transfer to a bowl. Wearing rubber gloves, seed, and finely chop the pepper. Finely chop enough onion to measure ¼ cup. Finely chop the cilantro. Stir the chopped pepper, onion, cilantro, and garlic into tomatoes with the lime juice, salt, and pepper. Salsa may be made 1 hour ahead and kept at cool room temperature.

Chicken Breasts 101: Grilled with Double-Fig Glaze

Free-range chicken is one of the most popular meats to grill—and the one that is most often destroyed when grilling. If you start with a high-quality, pastured, locally raised chicken and follow this simple method of salting the bird first, cooking it on the bone, and using indirect heat for the bulk of the cooking time, you will be amazed at the results: juicy, flavorful, tender chicken that tastes as chicken should.

Even though the breasts are the star here, this treatment works well for thighs.

serves 4

for the chicken breasts:

4 chicken breast halves, bone in, skin on
About ½ teaspoon kosher salt
Vegetable oil

Rinse chicken with cold water and thoroughly pat dry with paper towels. Sprinkle chicken with a good kosher or sea salt, taking care to work some under the skin. Cover and chill at least 2 hours and up to overnight.

Let chicken come to room temperature for 30 minutes before grilling. Meanwhile, prepare a gas or charcoal grill for indirect heat. For gas, turn all burners on high and close the lid. When temperature inside the grill reaches 400°F, turn off one burner; the area over the turned-off burner is the indirect heat section. For charcoal, light 4 to 5 dozen (48 to 60) briquettes and let them burn until covered with ash, about 30 minutes. Mound them on one side of the grill; the area over the section cleared of coals is the indirect heat section.

Using a Teflon brush or oil-soaked paper towel, brush the grill with vegetable oil. Place chicken, skin-side down, on the indirect heat section, and close the lid of the grill. Cook 15 minutes. Turn chicken over, close the lid, and cook another 10 minutes. Move chicken to direct heat section and cook, turning once, until skin is brown and crispy, about 5 minutes.

The chicken is done when a meat thermometer inserted into its center (avoid the bone) reads 160° to 165°F; the chicken will be slightly pink, but will finish cooking while it rests at least 10 minutes before serving. Serve hot, warm, or at room temperature topped with the fig glaze.

for the glaze:

makes about ¾ cup

- ⅓ cup extra-virgin olive oil
- 1 large shallot, minced
- 4 large fresh figs, diced and tossed lightly with brown sugar
- 4 dried figs, diced
- Zest and juice of 1 lemon
- ¼ cup water
- 1 teaspoon fresh lemon thyme, finely chopped
- Kosher salt and freshly ground pepper to taste

Place half the olive oil in a small skillet over medium heat. Add the shallots and cook for 1 minute. Add the fresh figs and turn when the first side has caramelized; now add the dried figs and after caramelizing them, remove the pan from the heat. Add the lemon zest, juice, water, and thyme to the same pan. Season with salt and pepper. The glaze will be very chunky and that's the idea.

Aromatic Buttermilk-Basted Chicken

Nothing speaks of old-timey tenderness like buttermilk! In fact, I believe a buttermilk revival is brewing. Although this recipe calls for chicken pieces, which are very convenient for a picnic or party, I have had great success using a whole chicken. For the entire bird, follow the same recipe but allow additional time for grilling.

If planning to entertain, brine the chicken up to 1 day before grilling. You can grill up to 1 day before serving, keeping chilled in an airtight container. Otherwise, keep the meat chilled (you can use an ice chest) until ready to grill. When transporting the chicken elsewhere, keep it in the brine and chilled.

serves 6

for the brine:

- 1 quart buttermilk
- 2 tablespoons finely chopped shallots
- 3 medium cloves garlic, minced
- 2 tablespoons kosher salt
- 2 tablespoons sugar
- 1 tablespoon freshly ground cumin
- 2 teaspoons chopped fresh oregano
- 1 teaspoon freshly ground pepper
- 4 medium chicken breasts, bone in
- 6 chicken thighs, about 2½ pounds total
- 6 chicken drumsticks, about 1¾ pounds total

In a large bowl, mix the buttermilk, shallots, garlic, salt, sugar, cumin, oregano, and pepper.

Submerge the chicken pieces in the buttermilk brine. Cover and chill for at least 4 hours or overnight.

Preheat the grill to medium-high.

Lift the chicken from the brine, pat dry, and discard the brine. Place the chicken pieces in a sturdy, greased roasting pan to use on the grill.

Place the pan on the grill directly over the flame and close the grill lid. Roast, turning the pieces after 20 minutes, and then move the chicken to the indirect heat side of the grill. Continue grilling over indirect heat for another 45 to 60 minutes, or until the chicken browns, is no longer pink at the bone, and a meat thermometer registers 160° to 165°F.

Serve hot or cold.

Brined Lime-Pineapple Chicken

In this recipe we are again using a brine that makes the pastured chicken parts especially tender. This fully flavored poultry pairs beautifully with the grilled pineapple and grilled lime halves.

serves 4

for the brine:

4 quarts cold water
¼ cup sugar
½ cup plus 1 teaspoon kosher salt, divided use
5½ to 6 pounds chicken parts with skin and bones

for the vinaigrette:

¼ cup fresh lime juice
2 tablespoons soy sauce
1 large garlic clove, minced
3 tablespoons finely chopped fresh mint
¼ cup finely chopped fresh cilantro
dash of Sriracha sauce (found in ethnic aisle of markets or Asian markets)
½ cup vegetable oil

for the fruit:

½ large pineapple, cut into wedges
2 medium limes, unpeeled, quartered lengthwise

to brine the chicken:

Bring water, sugar, and ½ cup salt to a boil in a heavy, 6- to 8-quart pot. Cool brine completely, then add chicken and soak, covered, and refrigerated 6 hours to overnight. Remove the chicken and pat dry.

to make the vinaigrette:

In a large bowl whisk together lime juice, soy sauce, garlic, mint, cilantro, Sriracha sauce, and the remaining teaspoon of salt in a large bowl, then slowly add the oil in a stream, whisking until combined.

Preheat the grill to medium-high.

Sear the chicken pieces in 2 batches on a lightly oiled grill rack, cover with grill lid, turn once until well-browned, and then move to the indirect flame of the grill. Grill about 30 to 40 minutes, turning occasionally to prevent charring, depending on the size of the chicken pieces. The chicken should read 160° to 165°F on a meat thermometer.

When chicken is almost done, grill the pineapple wedges and lime quarters (cut sides down), uncovered, until grill marks appear, about 3 minutes.

When finished, transfer the chicken to the bowl with the vinaigrette and marinate while fruit is grilling. The chicken is fine to serve at room temperature.

Remove chicken from vinaigrette and place on a serving platter with the grilled pineapple and lime and serve. Make sure to let your guests know to squeeze the lime over the chicken and pineapple.

Grill-Roasted Pear-and-Bacon Stuffed Chicken

One of our favorite treatments for a whole pastured chicken is to stuff and grill-roast the entire bird. For the first meal you have a very elegant dinner and then, if you're lucky enough to have meat left on the bones, you can use it for other scrumptious recipes.

serves 4

1 4-pound roasting chicken
Kosher salt and freshly ground pepper to taste
3 medium d'Anjou pears, peeled and cored, 1 diced, 2 coarsely chopped, divided use
1 tablespoon Dijon or spicy mustard
3 medium scallions, white and green parts, thinly sliced
6 ounces pastured bacon, cooked until crisp, drippings reserved
Fresh cornbread, cut into 1-inch cubes
2 teaspoons fresh, finely chopped sage

for the roasting aromatiques:

2 stalks celery, diced
1 cup apple cider

Preheat one side of the grill to medium-high.

Rinse the chicken in cold water and pat dry. Sprinkle the inside with salt and pepper.

for the stuffing:

Toss the diced pear with the mustard, scallion, bacon, cornbread cubes, and sage; then loosely pack the mixture into the chicken cavity. Tie the legs together with kitchen string and tuck the tips of the wings under so they don't burn.

Rub the outside of the chicken with a small amount of the reserved bacon drippings and season with salt and pepper. Place the chicken in the cast-iron roasting pan, breast-side down to begin grill-roasting. Sprinkle the reserved chopped pear and celery around the chicken in the pan. Cover the pan tightly with foil and grill over direct flame for 20 minutes. Turn, breast-side up, cover again with foil, and close the lid of the grill for 25 minutes more.

Remove the foil, pour the apple cider over the chicken and pears and finish grilling over indirect flame with the grill lid closed to brown the chicken until its juices run clear, and a meat thermometer inserted into the thickest part of a thigh reads 160° to 165°F; cooking time should be a total of 1 to 1¼ hours. Baste the chicken with the pan drippings every 20 minutes.

Remove the chicken from the grill and transfer to a serving platter. Arrange the fruit and vegetables around the meat and keep warm. On the grill, tip the roasting pan so the drippings pool to one end. Remove excess fat with a tablespoon. Scrape the bottom of the pan, add the remaining ¼ cup cider, and cook the juices for about 5 minutes. Drizzle the sauce over the stuffed chicken and serve more sauce on the side.

Sweet and Spicy Glazed Chicken

Here is another lovely treatment for grill-roasting a whole chicken, although this time it is only loosely stuffed with aromatics so it cooks in less time.

serves 4

for the glaze:

1 tablespoon soy sauce
2 tablespoons rice vinegar
½ teaspoon hot pepper sauce, or more to taste
⅓ cup orange liqueur
1 4½-pound chicken, fat and giblets removed, rinsed and patted dry
Kosher salt and freshly ground black pepper to taste
2 tablespoons, ¼ stick, butter, at room temperature
2 ½-inch slices fresh ginger, smashed and chopped
2 medium garlic cloves, smashed and chopped
½ orange, cut into 4 slices, skin on
½ cup water

Preheat the grill to medium-high.

to make the glaze:

Whisk soy sauce, rice vinegar, hot pepper sauce, and orange liqueur in a small bowl for the glaze.

Sprinkle the chicken cavity with salt and pepper. Loosen the chicken skin by running your fingers under it. Rub most of butter under the skin covering the breast and thighs; rub the remaining butter over the chicken. Place the chicken in a cast-iron skillet to put on the grill. In a small bowl combine the ginger and garlic. Squeeze some juice from each orange piece over the chicken, and then stuff the cavity with the orange pieces and the ginger-garlic mix. Tuck the wing tips under. Brush the entire chicken with the glaze, reserving some of the glaze.

Place the skillet with the chicken on the grill over direct flame, close the lid, and roast 20 minutes. Add ¼ cup of water to skillet, brush chicken with glaze, and roast 15 minutes longer.

Roast until a meat thermometer inserted into the thickest part of a thigh registers 160° to 165°F, brushing chicken with glaze every 10 minutes, about 40 minutes longer. Tilt the chicken to allow juices from its cavity to run into the skillet; let stand 10 minutes and then place on a serving platter.

Remove fat from the surface of the pan juices in the skillet. Add any remaining glaze to the pan juices in the skillet and bring to a boil; cook about 5 minutes. Serve this sauce on the side with the chicken.

Soft Shredded-Chicken Tacos

Everyone—young, old, and in-between—loves tacos as they are simple and satisfying. Soft tacos are becoming very popular. If you want to avoid the fried version, you can grill your own. Chicken breast or thigh meat will work with this recipe but the dark meat is more moist and it has a richer flavor. The components here are quite basic, but nothing beats a great taco!

If you want to make this meal more of a party, offer the toppings as a big spread and let people build their own. Include additional toppings, such as radishes, beans, fresh corn kernels, and olive slices.

serves 4

for the spice mix:

1 teaspoon ground oregano
1 teaspoon freshly ground cumin
½ teaspoon garlic powder
Kosher salt and freshly ground pepper
1 pound boneless, skinless chicken thighs or chicken breasts
12 6-inch white corn tortillas
1½ cups thinly sliced sturdy green lettuce such as romaine
¼ cup, about 1 ounce, queso fresco, artisan cheddar, or shredded Monterey Jack cheese
¼ cup sliced radishes
Avocado slices, sprinkled with lime juice to prevent browning

for serving:

sour cream and fresh lime wedges

Heat the grill to high.

In a small bowl, combine the oregano, ground cumin, garlic powder, salt, and pepper to make a rub; and press all over the chicken.

Grease the grill rack and then grill chicken over direct heat for 10 minutes on each side or until done. Let stand 5 minutes, remove bones, then shred (or pull apart) with your fingers.

Heat the tortillas on the grill (along with the chicken) and keep warm until the chicken is grilled. Divide the shredded chicken evenly among the tortillas; top the chicken with equal parts of the lettuce, cheese, and radishes. Add an avocado slice, if desired. Slather with pico de gallo, and red or tomatillo salsa with your choice of heat. Serve sour cream and lime wedges on the side.

Spanish Stuffed Chicken Packets with Romesco Sauce

When traveling in Spain, I learned to love romesco sauce and to appreciate the indigenous ingredients that comprise it. When these ingredients are featured in a recipe with pastured chicken, it somehow murmurs, "Ahh yes, Spain."

serves 4

for the chicken:

- 4 boneless, skinless chicken breast halves
- ½ cup Manchego cheese, finely shredded
- ¼ cup sweet onion, finely chopped
- Freshly ground pepper
- 2 tablespoons fresh Italian parsley, finely chopped

With a sharp knife, create a pocket in the chicken breast by making a slit lengthwise in the middle of the meat and cutting through toward the other side, leaving about ¼ inch whole and uncut.

Combine the cheese, onions, pepper, and parsley; divide evenly among the breasts and stuff the pockets.

for the romesco sauce:

- 2 tablespoons Marcona almonds
- 1 thin slice French bread, about 4 × 3 × ¼ inches
- 1 large garlic clove
- 1 roasted red bell pepper or 1 7-ounce jar roasted red peppers, drained
- 4 tablespoons olive oil, divided use
- 1 tablespoon sherry wine or red wine vinegar
- ¼ teaspoon cayenne

Heat the oven to 375°F.

Toast the almonds in the oven for just a few minutes to heat. Then toast the bread slices until golden crisp. Transfer the almonds to the bowl of a food processor. Tear bread into pieces and add to the processor. With the machine running, drop garlic through the feed tube and process until almonds and garlic are finely chopped. Add the red peppers, 3 tablespoons of the oil, vinegar, cayenne pepper, and process until mixture reaches the consistency of thick mayonnaise, scraping down sides of bowl occasionally. Cover and refrigerate. Sauce can be prepared 1 day ahead.

to grill the chicken breasts:

Heat one side of the grill to medium-high and place a medium cast-iron skillet over the fire. Heat skillet until hot.

Add the remaining 1 tablespoon olive oil to a skillet and heat. Add the chicken breasts and close the lid of the grill. Cook about 8 minutes until golden; turn and then cook 8 minutes more or until a meat thermometer registers 160° to 165°F. Remove the skillet and let chicken breasts rest for 5 minutes.

to serve:

Place the chicken on serving plates, drizzle with the pan juices; serve the romesco sauce on the side with a small amount topping the chicken.

Chicken Salad Niçoise

There are many variations on the Niçoise salad. Here we have a brilliant version—a gem from ingredients found in the Nice area of France. Julia Child brought the traditional dish to the United States and added lettuce leaves, which caused a good bit of controversy.

Our version includes many farm-fresh ingredients, which balances vitamin-rich vegetables with healthy protein-rich foods, such as seared chicken and eggs. The anchovies in the vinaigrette may be optional, but they do lend a subtle saltiness.

serves 2

8 new potatoes, about 2-inches in diameter, scrubbed, skin on
¾ pound green and wax beans, trimmed
3 baby crookneck squash or zucchini, halved lengthwise
2 ripe, medium Roma tomatoes, quartered
About 2 cups sliced, seared chicken breasts and thighs
½ cup Niçoise or small, brine-cured purple olives
2 hard-boiled eggs, quartered

for the vinaigrette:

3 tablespoons white wine vinegar
1 tablespoon Dijon or spicy mustard
1 small shallot, peeled and finely chopped
1 medium clove garlic, finely minced
2 tablespoons Italian parsley, finely chopped
2 tablespoons finely chopped fresh basil
2 fillets canned anchovies, finely chopped (optional)
Kosher salt and freshly ground black pepper to taste
½ cup of extra-virgin olive oil

Place potatoes in a pot of cold water and bring to a boil. Cook, uncovered, about 15 to 20 minutes or until tender. Remove potatoes with a slotted spoon and set aside to cool. Once cooled, cut into halves.

Add a little more warm water to the pot and bring to the boil again. Cook the green and wax beans until crisp-tender, about 3 to 4 minutes. Immediately transfer beans with a slotted spoon to a bowl of ice-cold water to stop the cooking. Use the same process for the squash, but barely cook, just 2 minutes.

to make the vinaigrette:

In a food processor add the vinegar, mustard, shallot, garlic, parsley, basil, anchovies, salt, and pepper and mix well. With the motor running slowly drizzle in the olive oil.

to assemble the salad:

Place the potatoes, beans, squash, tomatoes, chicken, and olives in a large mixing bowl. Pour on the vinaigrette and use tongs to gently toss the ingredients to coat well. Divide salad into 2 serving plates and place the egg on the sides.

Za'atar Chicken Wraps

Za'atar, an alluring spice blend used frequently in the Middle East, includes sesame seeds, thyme, sumac, and salt. Za'atar is sprinkled on vegetables and freshly baked flatbread or it can be mixed into olive oil or yogurt for a zippy dip. Spellings for this mixture are as varied as that of sumac, its main spice, so you may know this blend as zahtar, zither, or zatar.

Look for prepared za'atar in Middle Eastern markets, gourmet shops, and some mail-order sources.

serves 4

for the za'atar chicken:

1 medium clove garlic, finely minced
2 tablespoons za'atar
1 tablespoon extra-virgin olive oil
1 teaspoon kosher salt
1 to 1½ pounds boneless skinless chicken thigh or breast meat

for the wraps:

4 small square sheets soft lavash bread
2 cups thinly sliced crisp lettuce, such as Romaine hearts
½ medium English cucumber, thinly sliced
Spicy red chile sauce, such as harissa hot pepper sauce

for the tahini-yogurt sauce:

¼ cup tahini paste
1 cup plain yogurt
1 medium garlic clove, finely minced
1 teaspoon toasted sesame oil
1 teaspoon fresh mint, finely chopped
1 teaspoon honey
Kosher salt to taste

for the chicken:

Combine the garlic, za'atar, olive oil, and salt. Spread this mixture evenly over the chicken legs or breasts and then place the chicken on heavy aluminum foil.

Place the foil with the chicken in a hot smoker or on the indirect heat side of the grill and smoke for about 45 minutes or until chicken registers 160° to 165°F on a meat thermometer. Remove and let rest on a cutting board for 5 minutes, then thinly slice or chop, saving the juices.

for the sauce:

In a medium bowl, whisk together the tahini, yogurt, garlic, sesame oil, mint, honey, and salt until smooth.

to assemble the wraps:

Lay each lavash on a work surface. Place the chicken in juices on the left side of each leaving a 2-inch border on each side, then add tahini yogurt sauce, lettuce, cucumber, and a small amount of chile sauce. Roll up, tucking in the ends. Serve immediately.

Moroccan Chicken

Once I learned more about the colorful history of foods from Morocco and visited countries influenced by its cuisine, I was hooked. The use of fragrant spices is especially intriguing and unique in ethnic cuisines. Moroccan recipes don't have to be complicated, but they usually include cinnamon, which I prefer to keep to a pinch so it doesn't overwhelm.

serves 8

for the marinade:

1 cup extra-virgin olive oil
¼ cup balsamic vinegar
¼ cup fresh squeezed orange juice
3 medium cloves garlic, minced
2 tablespoons freshly toasted and ground cumin seed
1½ tablespoons freshly roasted and ground coriander seed
½ teaspoon ground cinnamon
2 teaspoons kosher salt
¼ teaspoon cayenne pepper
4 medium chicken breast halves with bone and skin
4 chicken legs
4 chicken thighs
¼ cup equal parts minced fresh Italian parsley and mint, for garnish

For the marinade, whisk first 9 ingredients in a large, glass baking dish. Add the chicken to the marinade; turn to coat. Cover with plastic wrap; chill 4 to 6 hours or overnight.

Heat grill to medium.

Remove chicken from the marinade and drain slightly. Reserve marinade for brushing chicken. Place chicken on barbecue and make sure to prevent flareups by moving away from flames temporarily. Grill chicken until just cooked through, occasionally brushing with any remaining marinade, about 10 minutes per side for breasts and about 12 minutes per side for leg and thigh pieces.

Transfer chicken to platter. Sprinkle with parsley and mint.

Pasture and Garden-Vegetable Kebabs

Everyone needs a basic chicken skewer recipe incorporating the fresh harvest of the season. The vegetables can be substituted depending on what is very fresh and available locally.

serves 8 as a snack or starter, 4 as a main course

4 boneless, skinless chicken breasts
12 firm button mushrooms, wiped clean with a damp paper towel
3 young firm crookneck squash (such as Zephyr variety, if available)
2 medium red or orange bell peppers, seeded and cut into 1-inch squares
2 small red onions, peeled and cut into 1-inch chunks
8 metal or wood skewers

for the marinade:

2 tablespoons extra-virgin olive oil
Zest and juice of 1 medium lemon
1 teaspoon chopped fresh tarragon or rosemary leaves
Kosher salt and freshly ground pepper

In a large nonreactive bowl, combine all ingredients. Use plenty of the salt and pepper.

for the chicken and veggies:

Cut chicken breasts into bite-sized chunks, toss in with marinade, and combine until thoroughly coated. Marinate for 45 minutes or overnight, refrigerated.

Soak the skewers in cold water. Mix the vegetables in with the chicken and marinate for another 30 minutes in the refrigerator.

Heat the grill to high.

Thread the chunks of chicken, and vegetables onto the soaked skewers. Place skewers on the hot grill, turning regularly, for 10 to 15 minutes, or until the chicken is golden brown and firm to the touch.

Serve immediately with additional fresh herbs sprinkled over the top.

TIPS FOR POULTRY STOCK THAT'S SIMPLE, DELICIOUS, AND NUTRITIOUS

The way I see it there are two different approaches to making stocks. The first is from the standpoint of a professional chef, who will transform ingredients into a "toney" and complex consommé. For the home cook, some basic ingredients are needed, such as aromatics in the way of bay leaf, peppercorns, or onions (of one kind or another). I don't recommend buying expensive leeks when an onion will do. It's all a bit of balancing and making the most of what you have on hand while creating a liquid base worthy of a soup or sauce.

Homemade stocks are great to have on hand. For convenience, make and store in the freezer for when you need it. When made from pastured chicken or turkey, a homemade stock far surpasses any commercial version of broth, which is what its weaker cousin is called.

Poultry stock is a great way to use the bones that are left when you debone breast or thigh meat. You can also make stock with a leftover cooked chicken or turkey carcass instead of fresh raw meat. It's best to use either chicken or turkey carcass and bones, but not a mixture of the chicken and turkey.

You can use fresh, whole vegetables or leftover vegetable trimmings and peels. I like to save the green parts of leeks or stems of parsley to use for stock. Do not use broccoli, cauliflower, vegetables of the cabbage family, bell peppers, or tomatoes because they overwhelm the stock. If you are not using roasted poultry and want a darker, richer stock, roast your raw poultry or poultry bones in a 450°F oven for about 20 minutes, before adding them to your stockpot. The caramelizing of the poultry bones and meat will yield a rich golden flavor.

Poultry Stock

Use an approximate amount of trimmings and leftovers instead of the whole vegetables listed here.

makes about one gallon

4 or 5 pounds of chicken or turkey parts or meaty poultry bones
3 whole medium bay leaves
1 whole leek, washed and cut in half, lengthwise
or
1 large onion, coarsely chopped
10 sprigs fresh Italian parsley with stems
2 or 3 large carrots, coarsely chopped
3 or 4 stalks celery with the leafy top parts, coarsely chopped
6 to 8 medium cloves garlic, coarsely chopped
1 tablespoon whole black peppercorns, cracked
1½ gallons cold water

Pour the water into a 12-quart stockpot and then add all your ingredients; Bring to a boil over high heat, then simmer on very low heat for about 2 to 3 hours. Occasionally skim off any foam that rises to the top.

Strain the liquid through a fine mesh strainer into another large heatproof container discarding the solids. Cool immediately.

When cool, place in the refrigerator overnight. Remove all solidified fat from the surface and store in a container with a lid in the refrigerator for 2 to 3 days or in the freezer for up to 3 months.

Goat

WHAT'S FOR DINNER IN MOST OF THE WORLD

Goat meat is the most widely consumed meat in the world, but you will have difficulty finding it for purchase in the United States except at specialty meat shops and ethnic markets. When you do find it, you may not see the typical cuts of meat found in supermarkets. Muslim tradition during the holidays requires that goats be cut into thirds: one-third for the family, one-third for friends, and one-third for the poor. Goats are ruminants, so grassfed goat meat provides all the nutritional benefits of other grassfed meats. Because goats are mainly browsers, their diet is more like deer in that they like eating the tender twigs of bushes and shrubs. They will also graze pastures, but goats do not feast on tin cans and T-shirts as is often thought—they are actually a bit finicky and very particular about having clean water to drink. The goat's browsing diet makes its meat very lean and gives it a flavor and texture somewhat like venison. Goat meat has less fat than lamb so it is not as tender as lamb nor is it as strongly flavored. However, if you have experience cooking venison, then you will quickly adapt to cooking goat meat. Because the meat is so lean, it does not require any of the dry aging that beef and, to a lesser extent, sheep need to tenderize the meat.

Raising goats is delightful and challenging. They can be quite endearing—and quite frustrating. One morning, as I was leaving the house, I saw the goats up on the roof of the farm's original icehouse (endearing); on another morning, the goats were in the strawberry patch enjoying the choicest berries (frustrating). According to Jeff Semler, a farmer friend, if you want to check your fence to see where goats will get through, just take a bucket of water and throw it against the fence. Wherever the water gets through, the goats will too (everywhere)!

About Goat Meat

The kid, called *cabrito* or baby goat, is especially tender. Even when young, however, goats have dark, strongly textured meat that does best when cooked slowly, a process that will tenderize the meat while bringing out its flavor. Cabrito is popular because goats are extremely hardy animals, capable not only of navigating difficult terrain, but of thriving in hostile environments like Africa, the American Southwest, Asia, and Latin America. The older goat, the *chevon*, has tougher meat and requires an even longer, slower cooking process to tenderize.

We used Kiko goat meat for these recipes. Kiko goats provide mild and quite tender meat despite being virtually fat free. Bred in New Zealand for meat production—named Kiko, Maori for *flesh* or *meat*—this breed of goat was brought to the United States in the early 1990s.

Due to its low fat content, cabrito meat loses moisture and toughens quickly when exposed to high cooking temperatures and dry cooking processes. Therefore, the two basic rules when cooking goat meat are:

1. Cook it slowly at a low temperature
2. Cook it with moisture

The tenderness of goat meat determines the method or methods best suited for cooking.

The tender cuts of goat meat are the legs, ribs, certain portions of the shoulder cut, the loin roast, and the breast. Less-tender cuts of goat are stew meat, riblets, and shanks. In general, it is advisable to cook the meat slowly. Cooking any meat at low temperatures results in a more tender and flavorful product with more juice.

Goat and Coconut-Vegetable Curry

Jeanne Dietz-Band of Many Rocks Farm in Maryland says it's important not to compare goat and lamb when cooking because goat is far leaner. This means that goat meat fares best at a low, slow heat, often within a braising glaze. In this recipe, the grilling is used to sear or caramelize the outside only, which leaves the interior rare. It will be finished off simmering in the rich, slightly spicy coconut curry sauce.

You may see quite a few recipes combining goat with a curry sauce: goat is the most widely consumed meat in the world, and each country has its own style of curry. We designed our Indian curry to respect the pastured younger goat meat that is worlds apart from the older chevon often seen roaming mountains. If the recipe seems too involved at first, don't turn away. The curry paste that forms the structure of the dish comes together quickly. As the tasters verified, it is a truly mahhhvelous cornucopia!

serves 8 to 10

for the curry paste:

2-inch piece ginger, peeled and chopped
1 medium onion, chopped
4 medium cloves garlic, peeled and smashed
1½ teaspoons turmeric
1 teaspoon kosher salt, or more to taste
1 tablespoon whole cumin seeds, toasted and finely ground
1 tablespoon freshly ground coriander seeds, toasted
2 mild green chiles, chopped, or to taste
2 pounds cubed goat meat
2 tablespoons vegetable oil

for the curry stew base:

2 bay leaves
5 green cardamom pods
2 14-ounce cans light coconut milk
1 quart, 32 ounces, chicken stock

In a blender or food processor, combine ginger, onion, garlic, turmeric, salt, cumin, coriander, and chiles until almost smooth.

for the meat:

In a medium bowl, coat the goat meat with half the curry blend and refrigerate overnight.

Heat the grill to medium-high, laying a flat griddle over the grill rack.

Grease the griddle lightly with the vegetable oil, then add the curry-coated goat meat cubes, spreading them so they don't touch. Sear quickly on all sides to brown, remove from the griddle, and place in a Dutch oven. To this pot add the remaining curry blend, the 2 bay leaves, cardamom pods, coconut milk, and chicken stock. Bring to a boil, reduce to a simmer, and cook about 1½ hours until the meat is very tender.

for the vegetables:

2 cups turnips, peeled and cut into 1-inch cubes
2 cups sweet potato, peeled and cut into 1-inch cubes
2 cups eggplant, cut into 1-inch cubes
3 cups butternut squash, peeled and cut into 1-inch cubes
2 large tart apples, diced
Lime wedges and cilantro leaves, for garnish
Rice, for serving

Meanwhile, as the meat cooks, add more oil to the griddle on the grill and add the vegetables. Close the lid, and grill until the vegetables are uniformly brown, occasionally tossing. Remove and reserve until the goat meat in the stock is tender.

Add the grilled vegetables to the meat, raise the heat to medium-high and cook another 10 minutes to warm the vegetables. Stir in the apples and finish cooking for 5 minutes. Serve on rice with the garnishes on the side.

Goat Cylinders with Citrus-Yogurt Dressing

Award-winning chef Pedro Matamoros of Matamoros Restaurant in Wheaton, Maryland, has developed close relationships with farmers in the Washington, D.C., area. His use of goat is inventive and showy because he forms the seasoned ground goat into a unique "skewer." We've used his basic recipe as inspiration and added a few of our own twists.

serves 12 as an hors d'oeuvre, 6 as an entree

for the skewers:

1 pound ground goat meat
Grated zest of 2 lemons, juice reserved for the citrus-yogurt dressing
1 teaspoon fresh rosemary, finely chopped
¼ cup pistachios, finely chopped
2 tablespoons dried figs, finely chopped
2 medium cloves garlic, finely minced
1 teaspoon sea or kosher salt
½ teaspoon freshly ground pepper
1 egg white, lightly beaten
Olive oil for grilling
12 skewers, soaked in water if wood

for the citrus-yogurt sauce:

1 cup thick Greek-style yogurt
1 tablespoon fresh mint, finely chopped
½ teaspoon kosher salt
1 teaspoon grated zest and ¼ cup juice of 1 orange
1 tablespoon reserved fresh lemon juice

In a medium bowl, mix the goat meat, lemon zest, rosemary, pistachios, figs, garlic, salt, pepper, and egg white. Mix well to make sure all ingredients are incorporated. Cover and refrigerate 1 to 2 hours.

Meanwhile, make the citrus-yogurt sauce by combining the ingredients in a small bowl. Refrigerate for 1 hour to let flavors meld.

Heat the grill to medium-high.

To assemble, shape meat mixture into cylinders, about 1-inch wide and 3-inches long. Insert the skewers lengthwise into each cylinder, brush with oil, and grill to sear on the exterior leaving the interior moist, about 4 minutes while turning to brown on all sides.

Place cylinders on serving platter and pass citrus-yogurt sauce.

To smoke goat leg bones

Make sure the bones have some meat left on them. You can use the smoked bones to make stock (page 173) and then pick the meat to use in soup (page 168). The small bone of the goat will not take too long to absorb the smoky essence. Keep in mind that too much smoke flavor will overwhelm the stock and, consequently, the soup. Put the bones on a sheet of heavy aluminum foil and place over the heat source. Close the grill lid for about 20 minutes to brown. Then reduce the heat to medium-low and move the bones and foil to the indirect side of the grill. Put some soaked fruitwood or herbs on another sheet of heavy foil to catch all the juices, and then place over the direct flame until smoke appears. Continue adding smoke fuel to the flames as needed and smoke for about 1 hour.

Greek Soup with Pistachio-Bruschetta

This is one of those double-use recipes: the meat, whether it is a goat leg with bone in or a leg of lamb, is first smoked for a meal. The meat that remains on the bone is used for the soup while the bone makes a deep, rich stock. We begin here using a stock that has already been prepared. This soup is a bit feisty from the heat in the Rotel tomatoes. If you prefer a milder version, simply use the very popular canned fire-roasted tomatoes.

serves 6 to 8

Put the bones on a sheet of heavy aluminum foil and place over the heat source. Close the lid for about 20 minutes to brown. Then reduce the heat to medium-low, move the bones and foil to the indirect side of the grill. Put some soaked fruitwood or herbs on another sheet of heavy foil to catch all the juices, and then place over the direct flame until smoke appears. Continue adding smoke fuel to the flames as needed and smoke for about 1 hour.

for the soup:

2 cups of the picked goat meat or lamb
2 tablespoons extra-virgin olive oil
1 large sweet onion, chopped
3 medium ribs celery, diced
3 medium carrots, grated
1 tablespoon flour
6 cups of stock (page 173) made from 3 pounds smoked goat bone or meaty lamb
1 teaspoon kosher salt
2 teaspoons dried oregano leaf
1 teaspoon freshly ground cumin from dry-roasted seeds
Zest from 1 medium lemon, save ½ teaspoon for the bruschetta
1 tablespoon fresh lemon juice
3 cups fresh greens cut into ribbons, Swiss chard, escarole, cabbage, or kale
1 10-ounce can Rotel tomatoes and green chiles
2 to 4 cups water, as needed
3 medium cloves garlic, minced

In a 10-quart stockpot, add the olive oil and heat over medium-high. When hot, add the onion, celery, carrots, and sauté for 5 minutes. Sprinkle in flour and allow to brown, stirring constantly. Pour in the stock then add the salt, oregano, cumin, lemon zest, and juice. Cook about 10 minutes, until the onion softens. Add the greens, tomatoes, and 1 cup of the water. Simmer, uncovered, for 40 minutes, stirring occasionally and adding more water until the soup achieves the desired thickness. Stir in the garlic during the last 5 minutes of cooking time.

Serve with a pistachio-bruschetta floating on top of each bowl of soup.

for the pistachio-bruschetta:

It's always fun to dress up a soup by adding an unexpected garnish. This tasty little morsel lends a fabulous texture to the soup in addition to that luscious essence of pistachio. You don't have to stop here with the bruschetta . . . it's a crunchy treat anytime. If you decide to go that route, be sure to double the recipe.

8 slices baguette, cut into 1/3-inch-wide slices
¼ cup extra-virgin olive oil for toasting the baguette
¼ cup natural pistachios, roasted
½ teaspoon reserved lemon zest
1 medium clove garlic, minced
¼ teaspoon toasted cumin seeds
½ teaspoon dried oregano leaf
3 tablespoons extra-virgin olive oil

Brush the slices of baguette with olive oil and toast until crisp.

In a food processor, add the pistachios, lemon zest, garlic, cumin seeds, and oregano and pulse the processor a few times until chunky. With the motor running, pour in the olive oil just to lightly mix but leave the pistachios with a bit of crunchy texture.

Kenyan Goat Kebabs (Mshikaki)

The popular skewered grilled goat meat, mshikaki, which actually stands for skewered grilled meat, is an East African classic. In the cities, vendors selling mshikaki are on every corner. Each one's mshikaki tastes different due to the terroir of the goat. Lamb can also be used in a recipe such as this along with a nice marinade. Make sure to trim off all fat.

serves 6

2 pounds young goat meat, cut into 1-inch cubes
2 limes, halved
½ cup extra-virgin olive oil
1 cup buttermilk
4 medium cloves garlic
1 teaspoon coriander seed, toasted and freshly ground
1 tablespoon cayenne pepper
White rice, cooked
Mango, cut into spears

Place the goat cubes in a large ceramic bowl. Squeeze limes and pour the juice over the goat. Place the olive oil, buttermilk, garlic, coriander, and cayenne into a blender and purée. Mince the garlic and add to the liquid along with the coriander and cayenne pepper. Pour this mixture over the meat in the large bowl. Cover, refrigerate, and marinate for 2 to 3 hours or overnight.

to grill:

Preheat the grill to high.

Skewer the meat and sear it over the hot grill for 15 minutes, turning to brown and basting with the remaining marinade.

to serve:

Place the skewers on a mound of rice and serve with the spears of mango and a tangy salad.

Goat Barbecue Beans with Bacon

Of course we are going to tell you that large, freshly cooked lima beans (sometimes known as butter beans) will yield some mighty fine beans, but if time or temperament doesn't allow, go for the canned variety. Fresh poblano chile peppers are fine to get to know because even when freshly cooked, they add a sweet smokiness. Many supermarkets now carry a good selection of chile peppers. The dish can be baked in a conventional oven or on the grill.

serves 6

8 ounces thick, nitrate-free sliced bacon, diced
1 large sweet onion, diced
½ cup fresh poblano chile peppers, diced
2 cups goat meat, grilled and shredded
4 cups cooked, large limas or 2 16-ounce cans butter beans, rinsed and drained
¼ cup dark molasses
¼ cup brown sugar
Few dashes hot sauce such as Tabasco, if desired
3 medium cloves garlic, minced
Kosher salt and freshly ground pepper to taste

Preheat the oven to 350°F.

Heat a medium-sized sauté pan over medium-high. When hot, add the bacon and sauté for 2 minutes to release some of the fat. Add the onion and poblano, and continue cooking for about 10 minutes. Drain any extra fat from the sautéed mixture.

In a large bowl, combine the shredded goat meat with the remaining ingredients. Finally, stir in the sautéed mixture and pour into a baking dish.

Bake uncovered at 350°F for 1 hour, stirring once after 30 minutes. Serve immediately or, equally as good, at room temperature.

Bourbon-Coffee Leg of Goat with Maple-Mustard-Horseradish Sauce

If you have spent any time around goats you know their legs are large and sturdy. A newborn kid seems to be all leg compared to its tiny body. The leg of a young goat will be smaller than the leg of a lamb so it really doesn't need a long grilling time. In this case, it is best eaten medium-rare to medium. The brine imparts a bit of sweetness and a hint of bourbon so the meat is mild and tender.

serves 4 to 6

for the brine:

1 3-pound leg of goat, bone in
Bourbon and Coffee Brine (page 209)

Make the brine and follow the directions to cover the meat with the brine. Let it work its magic in the refrigerator for at least 24 hours.

for the Maple-Mustard-Horseradish Sauce:

¾ cup grainy mustard
1 teaspoon bottled grated horseradish (spicy, if desired)
6 tablespoons maple syrup
2 tablespoons apple cider vinegar

Whisk all ingredients together in a bowl. This sauce can also be made ahead, kept covered and chilled, and brought to room temperature before serving.

to grill the leg of goat:

Preheat the grill to medium-high.

Remove the goat leg from the brine, pat dry, and place on a sheet of heavy foil. Place the goat leg on the foil directly over the flame and grill one side to brown for 10 minutes. Turn over and grill the second side for another 10 minutes. Reduce the heat to medium and continue cooking the meat until a meat thermometer registers 130°F for medium-rare, about 35 minutes total.

Let the goat leg rest, loosely tented with foil, for 8 minutes. Slice as you would a leg of lamb or separate the individual pieces of muscle (you can do this when cool enough to handle by running your fingers along each piece to separate). Then thinly slice each section across the grain of the meat.

Serve the goat meat with the sauce passed on the side.

Goat Stock

3 pounds smoked goat bones (page 167)
8 cups (2 quarts) water
2 bay leaves

Heat a medium-sized soup pot over medium-high and add the bay leaves and goat or lamb bones. When the water comes to a boil, reduce heat and simmer 2 hours to make a very rich stock.

At this point, cool the stock, remove the bones and meat with a slotted spoon. Remove any meat, chop it and save it for the soup (along with the stock). Discard the bones.

If making the stock for another use, cool and freeze the stock for up to 4 months.

All-Time Favorites: Burgers and Pizza

GIVING THE STANDARDS THEIR DUE

As much as we like to explore exciting new recipes, most of us have a special place in our hearts for standards, such as burgers and pizza. If you count them among your favorites, you're not alone. The average American consumes about 30 pounds of hamburger meat a year. And that's just the ground beef variety; if you include burgers made of ground turkey, chicken, and the all-vegetable types it would be even more. The amount of pizza consumed each day in the United States can be measured in acres—100 acres to be exact. That is the equivalent of 100 football fields.

From a culinary perspective, burgers and pizza are straightforward dishes. They're rustic, soul-satisfying, and they appeal to kids. Even better, they are foods that you can eat with your hands, which makes them convenient—anytime and anyplace—comfort foods.

There is nothing comparable to the smokiness of the grill and an open flame to bring out the seared, caramelized crunch of a burger holding a moist, juicy center or the crispness of a pizza crust oozing with melted cheese. We feel so strongly about these standards that we had to create a category to showcase them. Just start with the finest ingredients, a tamed flame, and let the sumptuous fun begin.

Wine and Bay Leaf-Infused Bison Burgers

You will want to take your time and savor these burgers. Wine with ripe, juicy fruitiness wraps itself around these grilled bison gems to set off fantastic fireworks of flavor on the palate. (Doesn't that sound like a double dose of terroir?) Because bison is low in fat, the added moisture infuses this clean-tasting meat with an intense flavor. If you can find fresh bay leaves you'll have a real winner. And guess what? You can also apply this same recipe to ground turkey or chicken; if you do, try it with white wine instead of red.

serves 6

2 pounds ground bison
1 tablespoon fresh lemon thyme, finely chopped
6 fresh bay leaves
8 medium cloves garlic, minced
2½ cups dry red wine
¼ cup fresh lemon juice
2 tablespoons balsamic vinegar
1 tablespoon extra-virgin olive oil
Kosher salt and freshly ground black pepper to taste

Mix the bison with the lemon thyme and form into 6 patties. Place the burgers in a flat, nonreactive baking dish with a bay leaf under each and garlic sprinkled over the tops.

Combine the wine, lemon juice, vinegar, and olive oil and pour over the patties. Season with salt and pepper. Cover and refrigerate, turning the burgers several times, for at least 3 hours.

Half an hour before cooking, remove the burgers from the refrigerator and drain on paper towels. Discard the bay leaves.

Heat the grill to medium-high.

Grill 3 to 6 minutes per side.

Serve on warmed rolls with fresh arugula and farm-ripe tomatoes.

Spicy Bison Sliders with Talbot Reserve Cheese and Curry Ketchup

Each of these sliders weighs about 1 ounce so the bison goes a long and tasty way as small bites or hors d'oeuvres. I have served these at my River Dinners and at a local cooking class because everyone loves a great burger. I topped each slider with a rich, cave-aged cheddar from Chapel Country Creamery in Easton, Maryland. You can garnish and build these sliders as you wish . . . adding to the bread or letting your guests add their own toppings, such as sliced onions, pickles, and lettuce or arugula. It's a good idea to make the curry ketchup in advance and keep it refrigerated.

serves 8

for the sliders:

12 ounces ground bison
Kosher salt and freshly ground pepper to taste
1 ciabatta loaf of bread, sliced horizontally, but not through, to open like a book
Freshly grated Talbot Reserve or other full-flavor, aged cheddar cheese, warm to room temperature

Heat the grill to medium-high.

Place the bison in a bowl and add salt and pepper. Form the meat into 2 rectangular burgers, ¾-inch thick, so they fit neatly onto the bread.

Grease the grill rack and add the burgers without crowding. Grill on one side until nicely browned, then turn and brown the other side, making sure that the burgers remain rare and juicy. While the second side browns, top the burgers with the cheese to allow it to melt. Place the opened ciabatta bread on the grill until just golden and warm. Then, while still warm, spread one side liberally with curry ketchup.

Remove the burgers to a cutting board and let rest for a moment. Lay the open ciabatta on the cutting board. Place each burger on the bottom half of the bread. Close the bread firmly and cut crosswise into 2-inch thick slices. A toothpick in each slice holds it all together and makes it easy to pick up.

for the curry ketchup:

1 12-ounce bottle chile sauce
1/3 cup sweet onion, finely chopped
1 teaspoon curry powder
1 teaspoon balsamic vinegar

Combine all.

Cheddar and Shiitake-Stuffed Burgers

Our friends at Boordy Vineyards in Hayes, Maryland, enjoy a gorgeous location and they make sure their vineyard supports the sustainable lifestyle. In the late afternoon on Thursdays during the growing season, they have a farmers market festival complete with farmers and their wares, live music, chef demonstrations, and festival food. Boordy supports our regional farmers and often they will hand out samples of their specially created recipes like the one they are sharing with us.

serves 4

¼ pound shiitake mushrooms, or other local variety, stems trimmed
3 small cloves garlic
½ medium onion, coarsely chopped
1 tablespoon unsalted butter
1 tablespoon extra-virgin olive oil
¼ cup Boordy Vineyards Shiraz or other Shiraz
¼ pound Keyes Creamery horseradish cheddar or other quality horseradish cheddar, shredded
Kosher salt and freshly ground pepper to taste
1 pound grassfed ground beef, bison, or turkey

Using a food processor, finely chop the mushrooms, garlic, and onion.

Heat a sauté pan over medium-high and add the butter and olive oil. When melted and bubbling, add the mushroom mixture and cook until soft, about 8 minutes. Add the Shiraz and cook 2 minutes longer or until the wine is absorbed. Cool. Add the cheddar and mix thoroughly.

Preheat the grill to medium-high.

Divide the ground meat into 4 portions. Divide each portion in half and flatten to make 2 patties. Season with salt and pepper.

Take 1 tablespoon of the mushroom and cheese mixture and place in the center of one patty; place another patty on top and pinch sides to contain stuffing and to make a burger.

Repeat for the second burger.

Grill the burgers over medium-high heat for 8 to 10 minutes each side and serve with favorite toppings.

Top-Notch Bison Burgers with Spicy Mayonnaise

Bison or beef will work in this recipe because they have similar properties, though the flavor will vary with the bison version tasting a bit earthier. If you are transporting this to an outdoor cooking event, the bacon and onion can be prepared ahead and brought to the event.

serves 4

1 pound ground bison or beef
2 tablespoons local honey
2 teaspoons kosher salt
2 teaspoons freshly ground pepper
4 strips of peppered, nitrate-free bacon
1 small red onion, finely diced
½ pound local Monterey Jack cheese, sliced

Heat the grill to medium-high.

In a large bowl, mix the ground bison, honey, salt and pepper; chill until ready to grill.

Place a cast-iron skillet on the grill rack and heat until hot enough to fry the bacon until almost crisp. Remove the bacon and set aside to cool, but leave the drippings in the skillet. Cook the diced onion in the bacon drippings until translucent, then let cool. Chop the bacon into ¼-inch pieces then add it, with the onions, to the seasoned bison, mixing well but lightly. Form into 8 very thin patties approximately ½ inch thick. Each thin patty will weigh about 2 ounces.

Fold the corners of 4 cheese slices toward their centers and place 1 each in the center of each of the 4 patties. Place the remaining 4 patties on top of these to make 4 large, thicker patties with cheese in the center. Shape patties as necessary.

Keep the heat on the grill, and add patties to the skillet. Cook until just medium-rare, about 4 minutes per side.

Moist grassfed burgers

Here's a great tip: If you're cooking grassfed beef hamburgers, add caramelized onions or other vegetables, such as fresh-grated carrots (or even fresh-grated apple) to add moisture and to compensate for less fat. Grassfed hamburger beef is generally 80 to 90 percent lean.

for the spicy mayonnaise:

¼ cup mayonnaise
1½ tablespoons Texas Pete or similar hot sauce

Combine and chill until ready to serve.

to serve:

1 large tomato, sliced
Romaine lettuce leaves
4 Kaiser rolls, split

Serve the burgers on Kaiser rolls garnished with the sliced tomato, romaine lettuce, remaining cheese slices, and the spicy mayonnaise on the side.

Treasure Burgers with Spicy Savory Pockets

This style burger began at the café I had in Santa Cruz, California, where I regularly grilled beef, turkey, and chicken over mesquite charcoal. The smoky aroma surrounded the café and floated down the block, which helped bring in the lunch crowd from the entire downtown area. You may have seen endless variations of this style of burger, probably because it always yields a moist, molten burst of flavor in the center. The inside pocket doesn't need extra cooking, so the burgers grill more quickly than a solid patty.

serves 4

1 pound ground beef, bison, turkey, or chicken
1 teaspoon Worcestershire sauce
2 tablespoons Dijon or spicy mustard
Lots of freshly ground pepper
¼ cup red onion, very finely sliced
4 tablespoons artisan goat or crumbled feta cheese (we used goat white cheddar containing chives), finely grated
Natural olive oil cooking spray or extra-virgin olive oil

Work the Worcestershire sauce into the meat being careful not to overmix. Shape the meat into 4 patties and make a deep well in the center of each patty.

In a small bowl, combine the mustard, pepper, onion, and cheese. Add this mixture, in equal portions, to the well in each patty. Lightly press the patties up and around to enclose the filling and smooth the meat. Coat the outside of the burger with the olive oil cooking spray or olive oil.

Heat grill to medium-high.

Grill about 5 minutes per side until golden brown, about 10 minutes total.

Treasure Burgers with Apple and Blue Cheese Pockets

Here's another variation on my Treasure Burger. In this recipe, the combination of blue cheese and apple imbue the burgers with satisfying bursts of savory and sweet.

serves 4

1 tablespoon unsalted butter, softened
2 tablespoons artisan blue cheese
¼ cup firm, fresh apple, grated
1 pound ground turkey or chicken
2 teaspoons fresh chives, finely chopped
Natural olive oil cooking spray or extra-virgin olive oil

In a small bowl, mix the butter, blue cheese, and apple.

Shape the meat into 4 patties and make a deep well in the center of each patty. Add the cheese mixture and chives. Lightly press the patties up and around to enclose the filling. Coat the outside of the burger with the olive oil cooking spray or olive oil.

Heat grill to medium-high.

Grill about 5 minutes per side until golden brown, about 10 minutes total.

THREE GRILLED PIZZAS

Pizza on the grill may be an entirely new venture for you, but it's a delightful skill to master. When you're done you will feel like you've taken a culinary standard and made something earthy and original out of it. Freeform and natural is the way to go—you don't want that production-line look with symmetrical crusts and mathematically spaced toppings.

Use these three pizza recipes as guides and then move on to designing your favorite combinations, styles, and even crust shapes. The easiest crust and probably the best choice for a simple test is the plain par-baked crust you find in the market, or take the time to play with a ready-made, raw dough ball. Our directions for making a basic pizza dough (page 191) lead you through the pizza dough-making process step-by-step. Eventually you will become such a pizza champ that making your own dough and grilling the pizza will be a breeze.

You can't deny that a grilled pizza, crusty and smoky from a live fire, is one of the most tantalizing dishes imaginable. The preparation can be complex, but with a little practice, it should soon become a more carefree experience.

As mentioned earlier, you have many options today for pizza dough. If you are a purist, or you just love the ceremony, make it from scratch from our recipe or use your favorite recipe. If not, some of the frozen pizza dough found in many markets tastes pretty close to homemade. Don't fret too much about the shape of the dough—hand-hewn and rustic is the most authentic. Of course, the simplest is the par-baked pizza shell that comes frozen or shelf-stable.

Once you're ready to make the pizza, begin by organizing your ingredients so the experience will be fun. Pizzas grill very quickly so you won't have time to regroup or find missing equipment. Set up a "prep station" or cooking table next to the grill where you can keep ingredients, tools, a cutting board to cut the finished pizza, and serving platters.

We suggest that you invest in a stainless-steel pizza peel (preferably one with a wooden handle) rather than one that is made entirely of wood. The care and maintenance is simpler with stainless and you won't have to worry about it catching fire.

Decide on the style of dough you want to use and then top it with one of our recipes or ad lib to create your designer-grilled masterpiece.

Pollo Pizza with Baby Bellas

serves 4

8 ounces grilled, pasture-raised chicken breast, shredded into bite-sized pieces
1 cup baby Portobello mushrooms, sliced ¼-inch-thick and tossed in extra-virgin olive oil
3 green onions, sliced
2 tablespoons extra-virgin olive oil plus extra for pizza dough
1 pound freshly made pizza dough or frozen pizza dough, thawed, and shaped into a desired pizza shell
2 medium Roma tomatoes, seeded and thinly sliced
1½ cups provolone, coarsely grated
8 fresh medium basil leaves, cut into fine strips

Preheat half the grill to medium-high.

Place the chicken breast strips, sliced mushrooms, and green onions in a medium bowl, and toss with the 2 tablespoons olive oil. Cover and let stand at room temperature for about 30 minutes to blend flavors. Prepare a space next to the grill for the toppings.

Brush the dough lightly with olive oil and flip, oiled-side down, onto the hot side of grill. Grill until marks form, about 2 minutes. Slide a pizza peel or spatula underneath to loosen the dough and to check for doneness. The dough may start to bubble at this point. Pop the bubbles with a fork or the edge of the spatula. Flip and transfer it to the indirect side of the grill (that is, no direct flame under the pizza).

Spread the tomato slices evenly over the pizza dough, sprinkle the chicken mixture evenly over the tomatoes, and then top with provolone.

Transfer the pizza back to the direct heat side of the grill. Cover the grill and cook until the cheese melts, about 4 to 6 minutes.

Transfer the grilled pizza to a cutting board, cut into pieces, and serve.

Sprinkle with the basil as you are serving.

Siracusa Pizza

This recipe from southeastern Sicily brings in a Greek influence with the feta cheese and olives.

As in all grilled pizza recipes, you may need to transfer the pizza back and forth from direct to indirect heat until you have conquered the balance. Here, balance means golden-brown dough, nicely melted cheese, warm toppings, and nothing seared dry. You are actually learning the artisan technique . . . you almost need to feel the cooking process.

serves 4

1 pound freshly made pizza dough or frozen pizza-crust dough, thawed
Extra-virgin olive oil
½ medium red onion, thinly sliced
6 ounces pork sausage, casing removed, formed into patties, grilled and then crumbled
½ teaspoon crushed red pepper (optional)
1 roasted medium red pepper, cut into ½-inch-wide strips
1 cup feta cheese, crumbled
½ cup pitted Kalamata olives, coarsely chopped
1 tablespoon fresh oregano, finely chopped

Heat half the grill to medium-high.

Have the pizza toppings prepared, measured, and placed on a tray or workspace. You will be building the pizza right on the dough over the grill and the "build" will go quickly.

Brush the dough lightly with olive oil and flip, oiled-side down, onto the hot side of grill. Grill until marks form, about 2 minutes. Slide a pizza peel or spatula underneath to loosen the dough and to check for doneness. The dough may start to bubble at this point. Pop the bubbles with a fork or the edge of the spatula. Flip and transfer the pizza to the indirect-heat side of grill (that is, no direct flame under the pizza).

Pizza dough and flatbread

Pizza dough can be made ahead and reserved in the fridge or frozen for up to two months. Flatbread has become synonymous with pizza. It was borrowed—and certainly rearranged—from the simple Italian version containing only flour, oil, lard, cheese, and herbs.

Top the dough with the onion, crumbled pork sausage, and sprinkle with red pepper. Spread the roasted red pepper strips evenly on top, and then sprinkle with the feta and olives.

Move the pizza back to direct heat for a few minutes, cover the grill, and cook until the bottom crust is golden. Now move it back to indirect heat, cover, and cook until the cheese melts, about 4 more minutes. Sprinkle the oregano over the pizza just before taking it off the grill.

Transfer the grilled pizza to a cutting board; cut into pieces, and serve.

Bok Choy Pizza with Beef or Pork

Bok choy, related to cabbage (the crucifer family), is loaded with nutrients and cancer-fighting antioxidants. It works beautifully roasted on top of a pizza. This delightful recipe can be used as an appetizer, lunch, or dinner. The olive oil-tossed bok choy adds a deliciously rich flavor to the ingredients and who doesn't love pizza, especially with the homemade touch? This is the fastest grilling of the pizza recipes because the pizza dough has been partially baked. For this version, you can build the pizza off the grill.

serves 4

1 12-inch par-baked pizza shell
2 tablespoons extra-virgin olive oil
3 medium garlic cloves, minced
8 ounces grilled rare beef or pork, cut into ½-inch-wide strips
8 ounces fresh mozzarella, thinly sliced
Mixed finely chopped fresh herbs of your choice
2 cups fresh bok choy ribbons tossed with extra-virgin olive oil
⅓ cup shredded Parmesan

Preheat the grill to medium-high.

Brush the oil and garlic onto the pizza shell. Sprinkle the meat strips over, follow with slices of the mozzarella, and sprinkle the herb mix over the cheese. Top with the bok choy and then the Parmesan.

Place the pizza directly over the heat for 2 to 4 minutes to brown the bottom. Move the pizza over to the indirect-heat side and close the lid. Continue grilling until the cheese melts and the crust crisps. Sprinkle with a bit more of the herb mixture. Cut, serve, and enjoy.

Basic DIY Pizza Dough

enough to make two 12-inch-round pizzas

1 cup warm water
2 tablespoons honey
½ teaspoon kosher salt
2 tablespoons instant-rise yeast
3½ cups all-purpose flour
¼ cup extra-virgin olive oil

Pour warm water into a bowl. The water should be about 85° to 115°F. Test it with your hand, but be very careful. It should feel very warm, but comfortable. Add the honey and salt. Mix by hand or with a wooden spoon until well blended. Add the yeast and mix some more. Let this mixture sit for about 5 minutes. Add 1 cup of the flour and olive oil and mix until well blended. Add the rest of the flour along with any other additions and mix well. The dough should turn into a ball. If the dough does not ball up because it's too dry, add water, 1 tablespoon at a time until it does. If your mixture is more like a batter, add flour 1 tablespoon at a time. Added water or flour may be needed to get the right consistency—just remember to do it in small amounts.

Once the dough balls up, place it on a floured board and knead for about 1 minute. This builds the gluten that helps the dough to rise. Place the dough in a lightly greased bowl covered with a damp, clean tea towel and store in a warm, dry area to rise.

After about 45 minutes, the dough should have about doubled in size. Punch down the dough and form a tight ball. Allow the dough to relax for a minute before rolling out.

to prepare crust and pizza:

Cut the ball in half and roll each half into a circle, square, or rectangle, ⅛-inch-thick, on an oiled cutting board. Brush dough lightly with olive oil and flip, oiled-side down, to hot side of grill. Grill until marks form, about 2 minutes. Flip and transfer to medium-heat side of grill.

10

10

Rubs, Mops, Marinades, Brines, and Sauces

Rather Classic Beef Marinade
Pork Marinade
Poultry Marinade
Dry-Roasted Seed Rub
Wine Brine
Spicy Wet Rub
Mustard-Madeira Sauce
Kansas City-Style Barbecue Sauce
Rita's Lacquer Sauce
Spicy Bumpy Barbecue Mop
Cherry Cabernet Sauce
Five-Spice Rub: Two Versions
Asian Five-Spice Rub
Fragrant Orange Asian Five-Spice Wet Rub
Double-Mushroom and Caramelized-Fennel Sauce
Bourbon Butternut Squash Chutney
Smoky Chunky Bacon Barbecue Sauce
Savory Lemon-Caramel Barbecue Sauce
Bourbon and Coffee Brine
Caramelized Onions
Juicy Smoked Tomatoes

MAKING FABULOUS MEATS TASTE EVEN BETTER

Moist, dusty, mild-mannered, devilishly hot, dense, or free-flowing, all of the following recipes add a complementary flavor without overwhelming the shining star—that pastured or grassfed meat. Consider all of the following recipes a crowning jewel that showcases the prized meat or poultry you are about to prepare—a veritable arsenal of flavor enhancers. Some of these recipes honor traditions, such as the Kansas City-Style Barbecue Sauce; others cross cultures with simple preparations that give hints of foreign flavors. Some recipes are just new ways of thinking about a choice ingredient, such as smoked tomatoes. A few of the recipes are simple yet good old commonsense basics with no-frills flavors.

DRY AND WET RUBS AND PASTES

A *rub* is a mixture of spices or herbs (or a combination of both), which is added to foods before cooking. Whether they are completely dry or they incorporate liquids, rubs are classified as either *wet rubs* or *pastes*. *Dry rubs* comprise dried ingredients instead of fresh flavorings—garlic and onion powders, for instance, are perfect. Rubs are most often used in barbecue and grilling because they stick to the meats to be grilled or smoked. In general, rubs start with paprika or chile powders (or both) to add color and mild flavor. From there, it is up to you and that's where the fun comes in! My pitmaster buddy, Jim Tabb, created his signature dry rub, called Pig Powder, which I use . . . *always*! The formula is said to be locked in a vault. Even his family does not know the ingredients.

Speaking of dry rubs, you may find that a mini grinder is indispensable for grinding and mixing small amounts of spice blends. Make sure the blades are very sharp. You also won't have a large food processor to wash.

Rubs seal in the juices and impart flavor. If you are working with them, here are a few tips to keep in mind:

- Rubs can be applied just before cooking. For a more pronounced flavor, apply rub to meat and then refrigerate for several hours.
- You can make your own rubs by combining your favorite fresh or dry herbs, spices, and other dry seasonings.
- To make a paste rub, combine dry seasonings with oil. Add small amounts of finely chopped garlic, onions, or wet seasonings, such as mustard, soy sauce, coffee, vinegars, or horseradish, to bind the mixture. Create a consistency that spreads thickly on your meat or poultry.
- The directions are the same for the beef, pork, and poultry marinades. Combine the ingredients, and then pour over the meat or poultry that has been placed in a nonreactive container. Marinate 2 hours or overnight.

Prepared rubs

If I happen to be in a rush and need to save a bit of time, I opt for some of my favorite prepared rubs, such as Jim Tabb's Pig Powder, which can be purchased in some grilling and specialty stores or found online. I also recommend Woodchicks (www.woodchicksbbq.com) but there are others that are also good, such as Vanns Spices (www.vannspices.com), the Kansas City BBQ Store (www.thekansascitybbqstore.com), and Signature Spices (www.signaturespices.com).

WHAT'S A MOP?

Mop is the fun name given to a sauce that is slathered over food before grilling and during the process. This friendly term describes the thickness of the sauce rather than any specific ingredient. Although a mop is a liquid that is thinner than a paste, it should not be runny or watery. A pourable barbecue sauce could be considered a mop.

MARINADES

A *marinade* is used instead of a dry rub when you want to add moisture to meats or poultry that have very little fat. The fats in marinades contain oil; brines do not.

Here are a few tips about cooking with marinades:

- Allow ¼ to ½ cup marinade for each 1 to 2 pounds of meat or poultry.
- When tenderizing, marinate at least 6 hours, but no more than 24 hours.
- Marinate tender cuts, such as tenderloin or round sirloin, for only 20 minutes to 2 hours to soak up the flavor.
- Marinate meat and poultry in the refrigerator, never at room temperature.
- Use food-safe plastic bags for marinating. A sturdy, glass baking dish is okay but remember that it is breakable.
- Turn or stir the meat occasionally to allow even exposure to the marinade.
- To use as a sauce later, save a portion of the marinade *before* adding the raw meat and make sure to prepare enough for both purposes. Marinade that has been in contact with uncooked meat *must* be brought to a full rolling boil for at least 1 minute before it can be used for basting or as a sauce.

Dry-Roasted Seed Rub

makes about 1/3 cup

- 1 tablespoon coriander seed
- 1 tablespoon sesame seed
- 2 teaspoons fennel seed
- 2 teaspoons ground cumin seed
- 1 tablespoon thyme, crushed
- 2 teaspoons kosher salt
- 1 teaspoon freshly ground pepper

Warm a small sauté pan over medium heat. Add the coriander, sesame, fennel, and cumin seeds and toast, swirling the pan constantly, until the seeds give off a rich aroma, about 1 minute. Immediately transfer to a plate and let the seeds cool.

Transfer the seeds to a mortar and pestle or a spice grinder. Add the remaining ingredients; grind to an even texture.

The rub is ready to use, or you can transfer it to a jar, cover tightly, and keep in a cool, dry pantry for up to 1 month.

Mix spices in a mini grinder.

Rather Classic Beef Marinade

makes about 3 cups

- 2 cups dry red wine
- ½ cup extra-virgin olive oil
- 1 tablespoon Worcestershire sauce
- 1 large onion, chopped
- ½ cup fresh Italian parsley, chopped
- 2 medium cloves garlic, minced or pressed
- 2 bay leaves
- 1 tablespoon fresh thyme leaves
- 1 teaspoon freshly ground pepper

Pour over the beef and refrigerate for at least 3 hours or overnight, turning several times.

Why roast whole seed and *then* grind?

When working with seeds, the rule is to *first* roast or toast the seeds and *then* grind just before using. No matter what process you use for grinding, the coatings or shells break to release the essential oils that give seeds their unique flavors. The resultant powder is best when fresh and protected against exposure to air, which is why we avoid commercially ground spices except for some dry rubs that will be used almost immediately. Spices, whole and ground, should always be stored in a cool, dark, dry place.

Pork Marinade

makes about 3 cups

½ cup peanut oil, no substitutions
¼ cup low-sodium soy sauce
¼ cup balsamic vinegar
Grated zest and juice of 1 medium orange
2 tablespoons Worcestershire sauce
2 tablespoons Dijon or spicy mustard
3 medium cloves garlic, minced
1 teaspoon freshly ground pepper

Poultry Marinade

makes about 1½ cups

¼ cup fresh lemon juice
¼ cup soy sauce
2 tablespoons unsulfured molasses
1 teaspoon Dijon or spicy mustard
1 tablespoon fresh ginger, grated
½ cup dry white wine

Wine Brine

You will note that there is no oil in this recipe, simply because it is not a marinade. Since we are incorporating a white wine here, this brine is best for light meats such as poultry or pork.

makes about 5 cups

1 bottle, 750 ml, Sauvignon Blanc, Pinot Gris, or other dry white wine
1 cup shallots, minced
1 cup fresh tarragon, chopped
¼ cup kosher salt
2 tablespoons sugar

Pour over the meat or poultry and turn to coat. Cover and chill for 1 day, turning once.

Spicy Wet Rub

This spicy wet rub can be toned down with a mild chile powder; give it more heat with an extra teaspoon or two of cayenne.

makes about ⅓ cup

2 tablespoons hot chile powder
1 teaspoon cayenne
1 teaspoon freshly ground pepper
2 teaspoons garlic powder
2 teaspoons fresh lemon juice

Store in the refrigerator in an airtight container for up to 2 weeks.

Mustard-Madeira Sauce

Madeira heightens this lovely sauce, rounding out and balancing its sweet, salty, tangy, and slightly spicy profile. This sauce is perfect for beef, lamb, or pork.

makes about 2 1/2 cups

- 2 tablespoons unsalted butter
- 2 small onions, minced
- 1 tablespoon Dijon or spicy mustard
- 1/3 cup Madeira or dry red wine
- 1¼ cups beef stock (page 74)
- Dash cayenne
- Kosher salt and freshly ground pepper

In a small saucepan over medium-high heat, melt the butter. When hot, add the onion and sauté until golden brown, about 4 minutes. Add the remaining ingredients and bring to a gentle boil. Let bubble until reduced by about 1/3 and thick and glossy.

Kansas City-Style Barbecue Sauce

Kansas City is renowned for its fine barbecue and this thick, sweet sauce is typical. To prevent the sauce from burning, brush it on the meat or poultry during the final minutes of cooking.

makes about 3 cups

¼ cup canola oil
2 large yellow onions, diced
4 large garlic cloves, minced
2 cups tomatoes, peeled, seeded, and diced
½ cup ketchup
½ cup firmly packed light brown sugar
¼ cup light corn syrup
2 tablespoons Worcestershire sauce
¼ cup molasses
½ cup apple cider vinegar
¼ cup Dijon or spicy mustard
½ teaspoon red pepper flakes
1 tablespoon paprika
Kosher salt and freshly ground pepper to taste

In a large saucepan over medium heat, warm the canola oil. Add the onions and garlic and sauté, stirring occasionally, until the onions caramelize and are very tender, about 12 to 15 minutes. Add all remaining ingredients except the salt and pepper. Stir to mix, bring to a simmer then reduce the heat to medium-low. Cook, stirring occasionally, until the sauce thickens, 45 minutes to 1 hour. Season with the salt and pepper.

Rita's Lacquer Sauce

My special Lacquer Sauce appears three times in this book because it works with so many dishes. The basic recipe is here but you will find variations on it in my Turkey Breast with Rita's Lacquer Sauce (page 123) and the Lacquered-Plus Pork Loin Chops (page 87), where it's given a little bit of a kick by adding fresh orange zest and orange juice. Take your own liberties with the base recipe to suit your tastes; add Sriracha sauce, lime juice, a bit of rice wine vinegar, or even Hoisin sauce. I am all about inspiring you to design your own version. Enjoy!

makes about 1/2 cup

1 tablespoon toasted sesame oil
1/3 cup apricot preserves
2 tablespoons low-sodium soy sauce
2 cloves fresh garlic, minced

In a small bowl, mix all ingredients together. Use part of the glaze to marinate and the remainder to top while grilling.

Warm a small sauté pan over medium heat. Add the cumin and coriander seeds and toast, swirling the pan constantly, until the seeds give off a rich aroma, about 1 minute. Immediately transfer to a plate and let the seeds cool.

Transfer the seeds to a mortar and pestle or a spice grinder. Add the remaining ingredients; grind to an even texture.

The rub is ready to use, or you can transfer it to a jar, cover tightly, and keep in a cool, dry pantry for up to 1 month.

Spicy Bumpy Barbecue Mop

When I manufactured specialty foods, my best-selling product was Bumpy Beer Mustard. Both men and women loved it, which made marketing it a real good time. Bumpy Beer Mustard also became the base for many a recipe, sauces in particular. In it I used three different kinds of mustard seeds. Sourcing those seeds may be difficult today but this recipe closely replicates the flavor.

makes about 1 cup

¾ cup spicy and sweet mustard, we prefer hot
2 tablespoons Worcestershire sauce
2 teaspoons Asian chile oil, which means hot
2 teaspoons white sesame seeds, toasted until golden

In a small bowl, mix the ingredients together and reserve until ready. When grilling, cook the food until 2/3 finished and then mop or spread over each side. When you're ready to serve, top the meat with more of the mop.

Cherry Cabernet Sauce

This fruity, upscale barbecue sauce marries beautifully with smoked meats and poultry. We tried it with our Hoisin Citrus-Tea Smoked Chicken (page 135). It is glorious slathered on pork. Make a full batch and keep it on hand in the fridge.

makes 3½ cups

1 tablespoon extra-virgin olive oil
1 medium onion, chopped
2 medium cloves garlic, minced
½ cup dry red wine
⅓ cup Hoisin sauce
⅔ cup dried tart cherries
½ cup chunky cherry preserves
3 tablespoons apple cider vinegar
3 tablespoons Worcestershire sauce
2 tablespoons brown sugar
2 tablespoons Dijon or spicy mustard
2 tablespoons fresh ginger, finely chopped
1 teaspoon freshly ground pepper
1 teaspoon toasted fennel seeds, freshly ground
¼ teaspoon cayenne
3 tablespoons fresh lemon juice, optional
¼ cup water, optional

Place olive oil into a medium saucepan over medium-high heat. Add onion and garlic and cook, stirring often, until limp, 3 to 4 minutes. Add remaining ingredients, except for the lemon juice. Bring to a boil then reduce heat and simmer, stirring occasionally, until liquid begins to thicken slightly, about 20 minutes. Let cool slightly.

You have a few choices here: you can leave the sauce chunky, which we prefer, or, for a smooth version, pour the mixture into a blender, add 2 tablespoons lemon juice, and blend until very smooth. Add a bit of water until you reach a thick but pourable consistency. Taste and add more lemon juice, if desired. Use warm or at room temperature.

Five-Spice Rub: Two Versions

Here you have a choice of a dry or wet Asian rub. Each of the rubs can be spread under the skin of poultry. These rubs work for any of the meats or poultry in this cookbook. In either case, simply mix the ingredients together.

Asian Five-Spice Rub

makes about 1/2 cup

3 tablespoons Chinese Five-Spice powder
1 tablespoon light brown sugar
1 tablespoon kosher salt
1 teaspoon cayenne pepper
½ teaspoon garlic powder

Fragrant Orange Asian Five-Spice Wet Rub

makes about 1/4 cup (can be easily doubled)

2 teaspoons extra-virgin olive oil
2 teaspoons freshly grated orange zest
2 teaspoons low-sodium soy sauce
2 teaspoons Chinese Five-Spice powder
1 teaspoon freshly ground ginger
¼ teaspoon freshly ground pepper
2 teaspoons honey

Double-Mushroom and Caramelized-Fennel Sauce

Beef, pork, and poultry can easily handle this vibrant caramelized vegetable sauce. I use a method here I named Steam-Sauté that accomplishes two cooking techniques in one pot. First, you are steaming to soften a dense food—fennel. Second, the small amount of added oil begins the sautéing after the water has evaporated. Normally, a thin layer of water, ¼ to ½ inch, will suffice in the bottom of a pan or skillet with a lid. I call it a layer of water because it steams effectively even with variables in the size of the pan. This efficient technique comes in handy to soften and caramelize, so keep it in mind.

makes about 3 cups

4 tablespoons extra-virgin olive oil, divided use
1½ cups baby Portabello mushrooms, thinly sliced
1½ cups shiitake mushrooms, stems removed and thinly sliced
2 cups fresh fennel bulb, thinly sliced
1 teaspoon fresh thyme, finely chopped
Kosher salt and freshly ground pepper to taste

Warm a medium-sized skillet over medium-high heat and add 2 tablespoons of the olive oil. When hot, add the mushrooms and sauté for 10 minutes until nicely browned. Remove from the skillet, and keep warm for a moment or two.

Place ½ inch of water in the bottom of the same skillet and bring to a boil. Add the remaining 2 tablespoons olive oil to the water. Add the fennel, cover, and steam for 8 minutes. Remove the lid, reduce heat to medium, and cook for 10 more minutes. Keep an eye on the vegetables here so they don't burn as the water evaporates. You may need to reduce the heat a bit. When the water has completely evaporated, and the fennel is tender and lightly browned, return the cooked mushrooms to the pan, add the thyme, and toss all to heat. Adjust seasonings by adding salt and pepper, and serve over grilled meat or poultry.

Bourbon Butternut Squash Chutney

This warm and lush chutney is very close to a product called Southern Bourbon Sweets, which I created and manufactured as part of a line of condiments. They are no longer available but this chutney comes close. The "Sweets" referred to sweet potatoes from North Carolina.

makes about 2 1/2 cups

½ cup apple cider vinegar
2 tablespoons yellow mustard seeds
¼ cup golden raisins
2 tablespoons extra-virgin olive oil
1 cup butternut squash, diced
½ cup apple, diced
½ cup onion, diced
¼ cup walnut pieces, lightly toasted
¼ cup bourbon

Bring the apple cider vinegar to a boil and pour over the mustard seeds and raisins, cover and steep for 20 minutes.

In a medium skillet, warm the olive oil over medium-high heat. Add the squash, apple, and onion, and sauté until tender, about 10 minutes. Mix in the walnuts, the raisin mixture, and the bourbon, and heat for about 5 minutes. Serve warm along with meat or poultry. This chutney will last well in the refrigerator for up to 3 weeks.

Smoky Chunky Bacon Barbecue Sauce

A small amount of bacon complete with drippings makes this sauce sing. The coarse texture is a surprise break in your repertoire of smooth barbecue sauces and it shouts homemade!

makes about 3 cups

- 2 slices applewood smoked, nitrate-free bacon, diced into ½-inch pieces
- 1 cup onion, finely chopped
- 1 small jalapeño, seeded and finely chopped
- 1 teaspoon freshly ground cumin
- 1 tablespoon canned chipotle in adobo sauce
- 2 cups very strong brewed coffee
- ¼ cup firmly packed brown sugar
- ¾ cup balsamic vinegar
- 1 cup tomato chile sauce (usually found in the ketchup aisle at the supermarket)
- 1 cup diced, canned fire-roasted tomatoes
- 1 tablespoon soy sauce
- 1 tablespoon Worcestershire sauce
- 2 medium cloves garlic, minced

In a 4-quart pot over medium-high heat, sauté the bacon until crisp.

Add the onion and jalapeño to the same pot and sauté until the onion is translucent, about 10 minutes.

Except for the garlic, add the remaining ingredients, and simmer slowly over low heat for 2 hours. As a grand finish, stir in the garlic and use immediately for grilling or let cool to store in the refrigerator for up to 6 months.

Savory Lemon-Caramel Barbecue Sauce

Just the name of this barbecue sauce makes me drool. It is rich, golden brown, and tangy, and a nice change from the ubiquitous red sauces. Use it on ribs, chicken, and just about everything. The Sriracha Chile Sauce included in this recipe might very well become your favorite spicy condiment. It adds character, not just heat (it's also called rooster sauce because many brands carry a rooster on their labels).

makes about 1¼ cups

½ cup light or dark brown sugar
1 teaspoon finely grated lemon zest
½ cup freshly squeezed lemon juice, or more to taste
1 tablespoon apple cider vinegar
2 tablespoons maple syrup
1 tablespoon Worcestershire sauce
½ teaspoon Sriracha chile sauce, found at Asian markets
1 teaspoon onion powder
½ teaspoon freshly ground pepper
½ cup Dijon or spicy mustard

Combine the first 9 ingredients in a medium, nonreactive saucepan over medium heat. Cook for 8 to 10 minutes, until a thick sauce forms. Add the mustard, taste, add pepper, and, if desired, more lemon juice, to taste. Transfer the sauce to a bowl or clean jar and let cool to room temperature. Cover and refrigerate until ready to use or store in the refrigerator for up to 2 months.

Bourbon and Coffee Brine

Because brines can simply be salt and water, it can seem excessive to add too many herbs and seasonings. I used to worry that the flavors would get lost in all that water. Well, this recipe changed my thinking. It not only tenderizes, it adds such richness you can almost taste the individual ingredients. I brined a whole chicken overnight and then smoked it, but the brine makes enough for any meat or poultry—grilled or smoked. As mentioned in the recipe, be sure to let the meat rest in the brine for 24 hours.

brines up to 6 to 8 pounds of meat or poultry

for the brine:

4 cups water
4 bay leaves
½ cup bourbon
1/3 cup kosher salt
½ cup firmly packed brown sugar
Zest and juice of 1 medium orange
4 medium cloves garlic, peeled and coarsely chopped
2 tablespoons black peppercorns
2 tablespoons Dijon or spicy mustard
2 tablespoons coriander seeds
2 cups strong coffee, cooled

Place 1 cup of water in a saucepan and bring to a boil. Add the bay leaves and let simmer for 15 minutes. Place in a large, nonaluminum brining receptacle; add the remaining 3 cups of water and remaining 9 ingredients and stir to mix. The brine should be cold at this point. Add the food to be brined, making sure it fully submerges. Refrigerate and let brine up to 24 hours.

Caramelized Onions

Think of a gently chilled batch of sweet caramelized onions as part of your repertoire—something you should always have on hand. The uses are more than I can count but here are a few ideas: they are terrific served as garnishes in soups, puréed as the base for a salad dressing or sauce, in sandwiches, or served as a side dish because just eating them as they are is a beautiful thing.

Farmers at the markets certainly know the sweetness of the onions they sell, but if you're going to a supermarket for onions to caramelize, look for Vidalia, Maui, Walla Walla, or Texas 1015s. However, with the huge, growing public appreciation for onions, it's likely you will be seeing more and more varieties available.

These edible bulbs are high in sugar content so they don't store well for much longer than a week at moderate room temperature. I sometimes store sweeter onions in the refrigerator.

The total cooking process for a large batch can take up to 45 minutes. The sugar content of the onions dictates the length of time needed. Patience is a virtue here.

makes about 3½ cups (includes butter)

4 tablespoons (½ stick) unsalted butter
3 pounds Vidalia or other sweet onions (4 to 5 onions, 3 to 4 inches in diameter), peeled and cut into ⅛- to ¼-inch-thick slices
Kosher salt and freshly ground pepper to taste
¼ to ½ cup water, broth, or dry white wine, optional

Melt butter in a large sauté pan over medium heat. When butter is foaming, add onions. Do not overcrowd.

Cook onions, stirring occasionally, until translucent, about 5 to 8 minutes. Continue cooking until deep golden-brown and caramelized, about 12 more minutes; add salt and pepper.

If the onions stick to the bottom of the pan, add a tiny amount of wine, broth or water to deglaze and loosen with a spatula. This liquid will (and should) evaporate. The entire process for a large batch can take up to 45 minutes, but doesn't need to. You be the judge.

Remove the onions from the pan and let cool.

for slower and slower caramelizing:

This is a gloriously simple way to get those juicy melt-in-your-mouth onions. It's a less tedious method if you need lots for a big batch of French Onion Soup. Increase the butter by 2 tablespoons to keep the onions moist.

Place the onions and butter in the insert of a slow cooker, cover, and cook on low for 8 to 12 hours, until the onions are deep golden-brown and very tender. The time will depend on the amount of onions you use and your slow cooker. If there is a lot of extra moisture after about 10 hours, turn the slow cooker to high to evaporate the liquid. It's almost impossible to overcook these; make sure to let the onions cook until they are mahogany colored and you can just see the richness.

Juicy Smoked Tomatoes

These tomatoes have been used in just about every way imaginable and always draw raves. In winter, they add a fresh blast of summer. A little goes a long way so just a smidgen in your marinara sauce will enliven. Using half of the smoked tomatoes in a salsa gives it remarkable character. I'm proud to say that Emeril Lagasse has become quite enamored with this recipe.

Choose a plum or Roma tomato, or any variety that has a high ratio of flesh to juice. I like Roma tomatoes because they are easy to find at farmers markets. Choose firm, ripe tomatoes. My version infuses the tomatoes with a bit of smoke, but leaves them moist and juicy.

to prepare the tomatoes:

Dice them into ½-inch cubes and drain in a colander for 30 minutes to remove the excess juice. Season the tomatoes with kosher salt and just a bit of sugar if they aren't naturally sweet. Place them on a greased sheet of heavy foil and then on the smoker grate or grill over indirect heat. These need to smoke at the low temperatures between 140° and 160°F.

Light smoke is all that's needed. The tomatoes are like sponges sopping up the smoke, and they can easily turn bitter if over-smoked. Check them every 20 minutes, rotating at least once. When finished, plunk them in a bath of extra-virgin olive oil to cover and add lots of freshly chopped garlic and the fresh herb of your choice. Rosemary works very well, but has an intense presence.

These tomatoes in the olive oil-garlic mixture will keep refrigerated for at least 2 weeks or frozen right in the olive oil.

Rita's indoor stovetop smoking method

Here in the Mid-Atlantic, the winters are fairly mild (2010 being a major exception) so I keep my grill active, even through a bit of snow. It always has been my modus operandi to grill year-round. However, with my love of smoky grilled foods, I have also adapted some indoor smoking techniques to create that deep, rich, earthy character without creating a room full of smoke. I used this indoor stovetop technique to test the Hoisin Citrus-Tea Smoked Chicken (page 135), but I originally developed the process for the Stovetop Smoked Tomatoes I prepared on an Emeril Lagasse television show.

My stovetop smoking method is excellent for tomatoes and perfect for a number of veggies, poultry, or seafood. I use a simple wok set-up, a small rack (close to the size found in a toaster oven), heavy aluminum foil, and aromatics, such as green herbs, rice, and white sugar (brown sugar would burn too quickly), for adding scent to the smoke.

for the wok:

In a small bowl, create the smoking mixture by combining a small amount of rice, tea leaves, and sugar.

Line the wok with a sheet of heavy foil (enough to fit inside the wok) and spray the foil with an olive oil cooking spray. Place the wok over high heat and add the smoking mixture, then add fruitwood leaves and sprigs; make sure not to make the pile too heavy because air circulation is necessary. Cover with a lid.

When small bursts of smoke begin to rise, place the rack holding the tomatoes or other food over the smoke source, cover with the foil—allow a small, gentle wisp of smoke to escape—and cook about 12 minutes on medium-high heat.

Remove the entire setup from the heat, but leave covered an additional 5 minutes or longer to infuse with smoky flavor.

Appendix A: Table Salt versus Kosher Salt and Sea Salt

It's no secret that salt enhances flavor. It brings out the character of ingredients and it can make everything come together. All salts are at least 97.5 percent sodium chloride; however, significant differences exist in the origin and processing of different salts.

Table salt, mined from underground salt deposits, has a small amount of added calcium silicate, an anticaking agent to prevent clumping. It has very fine crystals, a sharp taste, and, many believe, a somewhat bitter chemical flavor. Because of its fine grain, a single teaspoon of table salt contains more sodium than the same amount of kosher or sea salt.

Sea salt has become quite the boutique food item in the United States. You will hear chefs discuss the merits of the Hawaiian pink sea salt they sprinkle over the top of a grilled fish fillet. The Fleur de Sel (flower of salt), considered the very best because it is hand raked by French salt farmers, showcases the terroir of Brittany. Of course, retail and menu prices reflect this luxurious detail. Sea salt is harvested from evaporated seawater and receives little or no processing thus leaving intact the minerals from the water it came from. These minerals flavor and color the salt slightly.

Since high-end sea salts are usually expensive, keep in mind that they lose their unique flavor when cooked or dissolved. My friend, Ann Wilder, founder of Vanns Spices, taught me oceans about the world of salt and always advised using sea salt as a final garnish (and as a nice topic to chat about). I like to use sea salt for finishing a dish when the recipe directs cooks "to taste."

Kosher salt (also called "coarse salt") takes its name from the brining process—*koshering; to make kosher*—once used exclusively in Jewish practice for making meats and poultry kosher. Kosher salt contains no preservatives and comes from either seawater or underground sources (the packaging usually does not identify the salt's origin). The coarser grains, irregularly shaped crystal bits, dissolve beautifully and yield a robust saltiness. It is especially useful in brining, pickling, and preserving because its large crystals most effectively draw moisture out of meats and other foods; it is highly recommended for inclusion during the cooking itself. The price is fair and you know you're getting a quality product without additives or chemicals.

Appendix B: Farm Etiquette

FARM-VISIT ETIQUETTE: BEFORE YOU GO

Once you get interested in the benefits of local food, you will probably want to visit a farm. Please be sure to clear your visit with farm management; in other words, make an appointment or go on a day when the farm hosts guests.

AT THE FARM: BE COURTEOUS AND SAFE

- Be sure to check in and greet the farmer or other staff when you arrive.
- Always remember farms are workplaces: visit them with the same courtesy and respect you would show in a factory, school, or office.
- You will almost always see more than farm animals! Wildlife is abundant at farms: groundhogs, skunks, bees, mice, and snakes. There may be ample poison ivy, poison oak, sumac, and prickly multiflora rose (its hard thorns can slice the arm of an unwary passerby). If you venture into the woods and hedgerows, watch for holes in the ground. In other words, please make sure you and your guests know what to look for and how to take precautions against mishaps.
- Wear long pants to cover your legs; if you venture into high grass, check yourself and your guests for ticks.
- Take a hat and sunscreen, too. Most farms have acres of open, sunny areas.
- Take sturdy shoes that support you as you tramp through mud or climb over rocks.
- Take your own drinking water and, if appropriate, a brown-bag meal.

- Make sure that you take away any trash that you might create. Please don't expect the farm to handle it.
- Use bathroom facilities before your visit. If you're lucky, there might be a portable toilet available.
- If you want to take your dog, ask the farmer's permission first.
- Walk and speak softly.
- Get explicit permission to climb on any gates, walls, trees, stairs, hay, or equipment.
- Leave all gates exactly as you find them. If you are certain a gate is not as it should be, inform the farmer as soon as possible.
- Children should only enter a barn with an experienced guide.
- Many fences around pastures are electrified. People or animals that touch the fence will receive a jolt of electricity—not life threatening, but shocking nonetheless, so be careful.
- If walking through fields of produce or flowers, ask exactly where you may walk.
- If you are picking the produce yourself, thoroughly wash your hands after working in the fields; rinse your vegetables well.

ABOUT LIVESTOCK AND OTHER LIVING THINGS ON THE FARM

- If you stay calm, most animals will stay calm around you.
- Always get permission to feed or pet any animals.
- Chickens, ducks, turkeys, and guinea hens may range freely at a farm so find out about the farm's protocol regarding poultry.
- Don't turn your back on a rooster and *never* come between a rooster and his hens. He may attack. If that sounds amusing, it wasn't for the 3-year-old girl whom I saw attacked and her dress ripped.
- Don't run from a goose. It can get angry and chase you.
- Do not move behind the cows, where they cannot see you.
- Honeybees may live on the property. They are not aggressive unless you get close to the hives; however, if you are allergic to bee stings, you might want to bring an antidote for your own peace of mind.
- Ask if children may hold animals or pick up eggs. Make sure kids wash their hands afterward.
- Above all, let farmers know how much you value their work and that visiting their farms is a special treat. They appreciate acknowledgment of their hard work.

References

1. Durning A.B. "Fat of the Land." Washington, DC: World Watch Institute, 1991.
2. Cordain L. Grass-fed beef in the human diet: I. Historical and evolutionary significance. In *Proceedings of the National Grass-fed Beef Conference: The Art and Science of Grass-fed Beef Production and Marketing.* Harrisburg, PA: Feb 28–March 2, 2007.
3. Cordain L., Eaton S.B., Sebastian A., Mann N., Lindeberg S., Watkins, B.A., O'Keefe, J.H., Brand-Miller J. Origins and evolution of the Western diet: Health implications for the 21st century. *Am J Clin Nutr*. 2005;81:341–54.
4. Cordain L. Grass-fed beef in the human diet: II. Applications to clinical disease. In *Proceedings of the National Grass-fed Beef Conference: The Art and Science of Grass-fed Beef Production and Marketing.* Harrisburg, PA: Feb 28–March 2, 2007.
5. Duckett S. Benefits of Grass-fed Beef. In *Proceedings of the Mid-Atlantic Grass-fed Beef Conference*, Clemson University; August 2009.
6. Clancy K. "Greener Pastures: How Grass-fed Beef and Milk Contribute to Healthy Eating, Food, and Environment." Cambridge, MA: Union of Concerned Scientists, March 2006.
7. Kris-Etherton P.M. Polyunsaturated fatty acids in the food chain in the United States. *Am J Clin Nutr*. 2000;71(Suppl):79S–188S).
8. Daley C.A., Doyle P., Nader G., and Larson S. Added nutritional value of grass-fed meat products. In *Proceedings of the National Grass-fed Beef Conference: The Art and Science of Grass-fed Beef Production and Marketing.* Harrisburg, PA: Feb 28–March 2, 2007.

9. Duckett S. and Pavan E. Fatty Acid Profiles in Grass-Fed Beef and What They Mean. In *Proceedings of the National Grass-fed Beef Conference: The Art and Science of Grass-fed Beef Production and Marketing*. Harrisburg, PA: Feb 28–March 2, 2007.
10. Daley C.A., Abbott A., Doyle P.S., Nader G.A., Larson S. A review of fatty acid profiles and antioxidant content in grass-fed and grain-fed beef. *Nutrition Journal* 2010, 9:10. www.nutritionj.com/content/pdf/1475-2891-9-10.pdf
11. Russell J.B. and Rychlik J.L. Factors that alter rumen microbial ecology. *Science*. 2001; 292:1119–1122.
12. Christensen L.A. Soil, nutrient, and water management systems used in U.S. corn production. Agriculture Information Bulletin No. 774. Washington, DC: U.S. Department of Agriculture (USDA), April 2002. Economic Research Service (ERS).
13. Kaufman L. "Greening the Herds: A New Diet to Cap Gas." *New York Times*. 2009. www.nytimes.com/2009/06/05/us/05cows.html?_r=1
14. Ochterski J. Enhancing Pastures for Grassland Bird Habitat. Cornell University Cooperative Extension; November 2005.
15. Gardner B., Chase R., Haigh M., Lichtenberg E., Lynch L., Musser W., Parker D. Economic situation and prospects for Maryland agriculture. College Park: Center for Agricultural and Natural Resource Policy, University of Maryland; 2002.
16. Olsen J. 2004. A summary of basic costs and their impact on confinement vs. managed intensive rotational grazing (MIRG). Madison: University of Wisconsin Center for Dairy Profitability; 2004.
17. A.T. Diplock1, J.-L. Charleux, G. Crozier-Willi, F.J. Kok, C. Rice-Evans, M. Roberfroid, W. Stahl, J. Vina-Ribes. Functional food science and defence against reactive oxidative species, *British Journal of Nutrition* 1998, 80, Suppl. 1, S77–S112.

Resources

Without the gracious donations of these generous pasture-based farmers, our cookbook would not have been possible. Working with these exceptional farmers made our job even more exciting and delicious. Please know you are a big piece of our book.

Burgundy Pasture Beef
Grandview, Texas
www.burgundypasturebeef.com

Drover Hill Farm
Earlville, New York
www.droverhillfarm.com

Evermore Farm
Westminster, Maryland
www.evermorefarm.com

Grassland Beef/U.S. Wellness Meats
Monticello, Missouri
www.grasslandbeef.com

Gunpowder Bison and Trading
Monkton, Maryland
www.gunpowderbison.com

Hardwick Beef
Hardwick, Massachusetts
www.hardwickbeef.com

Jamison Farm
Latrobe, Pennsylvania
www.jamisonfarm.com

Lindenhof Farm
Kirkwood, Pennsylvania
www.lindenhoffarm.net

Many Rocks Farm
Keedysville, Maryland
www.manyrocksfarm.com

Polyface Inc.
Swoope, Virgnia
www.polyfacefarms.com

Index

Note: Bold page numbers indicate recipes